DISMANTLINGS

DISMANTLINGS

WORDS AGAINST MACHINES IN THE AMERICAN LONG SEVENTIES

MATT TIERNEY

CORNELL UNIVERSITY PRESS

Ithaca and London

First published 2019 by Cornell University Press

Printed in the United States of America

Library of Congress Cataloging-in-Publication Data

Names: Tierney, Matt, 1976– author.
Title: Dismantlings : words against machines in the American long seventies / Matt Tierney.
Description: Ithaca, New York : Cornell University Press, 2019. | Includes bibliographical references and index.
Identifiers: LCCN 2019020281 (print) | LCCN 2019021276 (ebook) | ISBN 9781501746772 (epub / mobi) | ISBN 9781501746567 (pdf) | ISBN 9781501746413 | ISBN 9781501746413 (cloth : alk. paper)
Subjects: LCSH: United States—Civilization—1970– | Technology—Social aspects—United States—History— 20th century. | Radicalism—United States—History— 20th century. | American literature—20th century— History and criticism. | Technology in literature.
Classification: LCC E169.12 (ebook) | LCC E169.12 .T557 2019 (print) | DDC 973.92—dc23
LC record available at https: / / lccn.loc.gov / 2019020281

For students, again

examine the heart of those machines you hate
before you discard them
and never mourn the lack of their power
— Audre Lorde (1973)

technology is freedom's other name when
— a meadowlark
comes sailing across my windshield
— Carter Revard (1976)

CONTENTS

DISMANTLINGS

Introduction
For the Sake of Survival

It was the computer engineer Alice Mary Hilton who coined the word *cyberculture*. In her only book—*Logic, Computing Machines, and Automation* (1963)—cyberculture inaugurated a forked path of technological ethics: in one direction lay unremitting innovation; in the other lay peace and productive work. Her formula was succinct:

machines for HUMAN BEINGS,
or human beings for THE MACHINE[1]

In this book, with Hilton's dichotomy in mind, I map critical and literary tools for prying apart presumptions about the centrality of technology to culture. What can literary thinking do, and what has it tried to do, to enrich and enable the emancipatory critique of technology? What has spoken and written language contributed to the transformation or destruction of reactionary institutions and ideas? My answer is words against machines; or rather, words against the forms of exploitation identified with machines, or with some machines but not all, or with machinic thought and the becoming-machine of laboring bodies. I propose the word *dismantling* to describe a way of reading with roots in a particular period of literary and theoretical production, the decade and a half before 1980: what I, following historians of labor, call the Long Seventies. In those years the radical imagination took unique shape through speculative thought and literary experiment. Central to this formation was a critique of technology, by which dreamed-up words came to take aim at real machines. This book is a study of some of those words. By asking what literature and activism said to technological thinking during a time when all three were in dramatic transition, I argue for a theory of technology and an attitude toward literature that will be suited to the task of critique in this digital and reactionary moment.

The time of this writing may be one of the rare moments when technology becomes visible for what it is: less a precise ontological condition, and more a loose collection of machines and an increasingly coherent set of political and cultural imperatives that lend themselves to wealth consolidation and state power. In a time of newly popular skepticism toward computation, an opportunity exists to study the legitimating language of the cultures of technology, while developing or recovering a contrary language that would be sufficiently critical of those cultures. When technical knowledge is transferred largely through procedures of updating and onboarding, or through vocational training in computation, digital technology is assumed as a component in institutions of knowledge and commerce. Yet at the same time, when the business of computation has been laid bare in its cravenness, and hashtags from #BoycottAmazon to #DeleteFacebook proliferate across social media, it can feel like the criticism of technology is obvious, even fashionable. Technophilia on one hand, technophobia on the other.

But the opposition of technophilia to technophobia is just a dodge and a distraction. These terms do not only impede an understanding of technology. More important, they impede forms of group organization that, moving beyond mere understanding, might actually embrace or invent other kinds of machines while smashing or relinquishing the worst existing ones. When both the technophilic and the technophobic attitudes fall away, and when no conciliatory middle path opens up between them, another vocabulary presents itself. This other vocabulary reflects another way of thinking. It leads away from the presumption that technology is a main driver of either oppression or betterment, while equally eschewing metaphors of the world that would regard technology as a neutral metal artifact that is necessary to human development. It leads not only away from accepted ways of talking about technology, but also away from visions of the world as a plunderable resource and a geographic expanse, a problem of access and distance, to be traversed by messages and travelers and goods at ever-greater speeds. It aims toward imaginative and conceptual tools in the literature of the recent past, new words now, through which to imagine something other than instrumental thinking in an exploitable world.

The present book is a counterlexicon drawn from a prior moment when the effects of technology appeared visible, even obvious, and poetry and fiction had something to say about that. When I call this book a counterlexicon, what I mean is that it excavates and defines seven of dismantling's forms of appearance: Luddism, communion, cyberculture, distortion, revolutionary suicide, liberation technology, and thanatopography. In these guises, dismantling is an exercise in, as well as an object of, what Siegfried Zielinski has called media-implicit analysis. For Zielinski, in his polemic *After the Media*, there is a way to understand technologies and their effects without deferring to their purported revolutionary capacity. Dismantling, like the practices that Zielinski describes, is a critical and cultural practice that is media-implicit rather than media-explicit. It is not about the revolutions that technology makes, but instead about the place of technology critique in practices of revolutionary thought and action. Like Zielinski's book, then, *Dismantlings* explores how technical "phenomena are integrated as subjects of research in wider discourses or epistemes . . . such as history, sexuality, subjectivity, or the arts"; and it stands apart from media-explicit discourses in which "individual media or a random collection of media or the media in the strategic generalization *expressis verbis* are the subject of [the] exposition."[2]

Dismantling thus involves an "exact philology of precise things"— concerned less with ontologizing particular technologies than with refining terms in the vocabulary of technology critique.[3] What are "precise" are the discursive and poetic constructions in the linked histories of technology, race, gender, violence, community, and responsibility in the world. Such constructions become visible through an experimental perspective, writes Zielinski, that is oriented toward "understanding the past not as a collection of retrievable facts but as a collection of possibilities."[4] Not intended to generate a new set of master terms, or to impose a new master narrative of the vanished past, dismantling is instead such a collection of possibilities. Constellated in this way, as an idea made material by many parts in motion, dismantling adds to a language of technology and politics that is both literary and contrary. Like other constellations, the word by which Walter Benjamin defines ideas insofar as "they are neither their concepts nor their laws," dismantling does "not serve the knowledge of phenomena." Rather, it is a varied means of technology critique, irreducible to the machinic phenomena that it considers, by which "inasmuch as the elements are grasped as points in such constellations, the phenomena are simultaneously divided out and saved."[5] Dismantling is a way to speak and write, as well to learn and move beyond.

My reference point for dismantling is the poet and essayist Audre Lorde, who famously demanded a coalitional politics in her 1979 argument that "the master's tools will never dismantle the master's house."[6] The 1970s, so often dismissed as an ethico-political wasteland between the disappointments of the 1960s and the terrors of the 1980s, was in fact a moment when a transformative disciplinary and cultural politics still seemed possible. Moreover, in the same moment, there was an unusual convergence of literary and political concerns with a radical critique of computation and telecommunication. This book accepts Hilton's dictum as the first utterance of a historical phase that ends with Lorde's own dictum. The Long Seventies proceeds, that is, from "machines for HUMAN BEINGS" to the dismantling of the master's house. This is a periodization of convenience, intended less to rearrange events of the past, and more to disrupt the artificial decade-by-decade regimen by which the past is often recounted, and to isolate a theoretical problematic: words against machines.

Yet the periodization is not just a convenience. As an event in cultural and intellectual history, Lorde's dictum does name something of the revolutionary imperative to invent by destroying, just as Hilton does begin something with her 1963 coinage of *cyberculture*:

> The changes that are being brought about by the cybercultural revolution—the automation revolution—will be of such staggering proportions to make the changes brought about by the first industrial revolution seem minute in comparison. Such changes are not remote from the concerns of the individual. . . . How will he survive in a world where standardization is becoming more widespread every day, and individual privacy less sacred? . . . Can we learn to use our marvelous new machines for the benefit of mankind without destroying in ourselves the very essence of being human? These are serious questions we must all ask ourselves. They demand of us an immense awareness of the social dangers in the wake of our new technology.[7]

I return in chapter 3 to Hilton's foundational call for an "awareness of the social dangers" of new technology. For now, it is enough to say that at the dawn of the Johnson administration, there issues an echoing call to think vigorously about machines. In the same period that soon brings both the Vietnam War and the Civil Rights Act, there is this startled realization: the war machine, the industrial machine, the computer, and the mechanisms of state are all connected. Built from the same parts, they can therefore be destroyed or overhauled by the same means.

One and a half decades later, calling for new coalitional movements on the precipice of the Reagan era and the leading edge of the era of home computers, Lorde will issue a poetically Luddite plea to dismantle the master's house. The arc that leads from Hilton to Lorde is not the period of affirmation in autonomous social and cultural factions, or of a simultaneous but separate revolution in computation, that the seventies are often taken to be. Instead, the period is better characterized by dismantling; by the literary and political contestation of technology and its metaphors; by an increasingly refined commitment to a technopolitical critique; by a growing dedication to revolutionary cross-movement coalitions; by commitments to breaking apart old systems of power, to refusing the inevitability of their replacement by new systems that would be just as exploitative, and to developing new communal practices in the wake of that refusal. This is the paradigm of literary and activist thought that is worth retrieving for the present.

When Sara Ahmed asks now how feminists and antiracists can build critical spaces in this present, her answer is still dismantling. She writes, in a way that resonates through the present book: "We need to dismantle what has already been assembled; we need to ask what it is we are against, what it is we are for, knowing full well that this *we* is not a foundation but what we are working toward."[8] Dismantling has such value for contemporary political thought, as it has in these words by Ahmed, because the Long Seventies and the present moment have a lot in common. Most significantly, both moments are seized with a critical discourse about technology, and by a popular social upheaval in which new social movements emerge, grow, and proliferate. This coincidence of the two moments may be chance, or it may be owed partly to the fact that many of today's most dire economic trends (like the move toward part-time work and low-wage precarity) began in the Long Seventies.

The Long Seventies were, in the words of Mike Davis, "the last great period of mass mobilization."[9] Yet they also began a spiraling downturn, tangible to many on the Left, toward a catastrophe that would be simultaneously ecological, technological, and human. It should therefore also be admitted, with labor historian Lane Windham, that "the 'long 1970s' did indeed prove to be an economic turning point that set the stage for working people's present crisis. . . . It was the birth of a new economic divide."[10] Yet if the Long Seventies are what initiated present forms of inequality, they may yet also offer the most valuable possible response. While there may not have been any clear victories in the feminist, antiracist, or environmentalist struggle of those years, they remain, in the words of the historian Cal Winslow, "strike prone years" in which "rank-and-file workers led wildcat strikes, rejected contracts, and forced official strikes . . . [alongside] the other protest movements of the 1960s and '70s—the black and women's movements, the anti-Vietnam movement, and the student movement, each of which profoundly influenced the workers' rebellion."[11] If we are to generate mass mobilization today, it follows that today's movements should continue to learn from the previous mass mobilization, not as a source of nostalgia, but instead as a model of imagination and planning. The youth movements and workers' movements of today must feed off one another, and off movements toward racial and gender equity, while incorporating the critiques of technology and technopolitics that are as urgent now as they were forty and fifty years ago.

Sociocultural criticism of technology largely sputtered to a halt around 1980, even as the media-explicit fields of communication studies and philosophy of technology approached their maturity. There are several reasons for the setback for radical cultural and political struggle around that time, and for the concurrent decline in the radical critique of technology. Home computers and early gaming systems became features of many homes beginning in those years, just as automation and nuclear power had already become fixtures of militarization and industry. Advanced telecommunications, and the attendant metaphors of connectivity and connectability, seemed at the same time and by increasing consensus to have defined the human experience. Criticizing the machines and their metaphors went from a popular standpoint to a minoritarian one. Simply put, it got harder to criticize the technoculture without sounding like a throwback or a romantic.

Meanwhile, the 1980 presidential election could offer the Left only a series of nested disappointments (as Ted Kennedy lost a primary to the centrist president Jimmy Carter, who lost in the general election to Ronald Reagan, an ex-governor and ex–movie star, widely regarded as a populist empty suit). The cultural historians Howard Brick and Christopher Phelps note: "A return to left-wing themes of economic inequality, working-class life, and struggle, capitalist instability, and democratic planning during the 1970s coexisted with the growing 'new' movements." And yet, as Brick and Phelps conclude: "Efforts to build a next left in the 1970s were, however, ultimately unavailing, whether envisaged in revolutionary, radical populist, or social-democratic terms. Even at a time when confidence in capitalist normality seemed to crack in a period of crisis, a reinvigorated ascendant political New Right surged ahead."[12] The 1980s, like the moment of this writing, mark an ostensible defeat to social movements; and the terms of that defeat were as visible after Reagan's election as they are now. What Brick and Phelps consider a "coexistence" of social movements in the 1970s is what I consider a coalitional approach that flowers in those years, having extended from roots in the middle 1960s toward an uncoordinated, multivocal discourse about literature and technology. Literary expression and technology criticism "coexist" alongside, and in reciprocal relation to, the movements in racial, sexual, environmental, and economic justice.

The deeply felt defeat of leftist coalitions at the dusk of this period was described in 1980 by Marxist literary critic H. Bruce Franklin:

> By the end of the 1970s, both double-digit inflation and massive unemployment had become chronic, the figures for both the national debt and consumer debt had become virtually astronomical, the per capita ratio of citizens in prison had become the highest in the world and was increasing each year, and the most common solution being offered to everybody's problems was some kind of drug. . . . Although the movements for radical social transformation launched in the 1960s had vastly extended their social base, they were far less organized than the powerful forces of reaction, and no resolution for any of the major social confrontations seemed in sight.[13]

Things fell apart. But they did not fall apart entirely or uniformly. As Franklin notes, even as reactionary forces overpowered radical forces, the latter expanded greatly in those years. This expansion is the tendency toward coalition and coexistence. And particularly as it veers from political movements into literary culture, this expansion involves the development of tools for a critical response to reactionary ideas and applications of technology. This is what dismantling is, and it has two primary meanings. In common use, to dismantle means to take apart, to break a thing into its separable pieces. In its etymology, to dismantle means to uncloak, to remove from a thing its mask or defenses. When a house is dismantled, its floors are torn up, exposing the ground for the construction of something else, something wholly new. When that house is the master's house, its dismantling is also a political triumph and a form of communal display. It is a way for activist thinkers to show each other how power was built and maintained in the first place. As the master's house is taken apart, so it is also exposed. As it is deconstructed, so it is also demythologized. As the ground is cleared, so too are foundations of power made visible.

Cyberculture

In 1946, Norbert Wiener wrote an open letter, published early the next year in *The Atlantic*, to a military research scientist at Boeing who had requested access to Wiener's research. Wiener, founder of cybernetics, cautioned that the invention of atomic bombs and concentration camps had signaled a change in which new social obligations must now accompany the communication of scientific knowledge. A year after World War II, as he was writing his field-defining book *Cybernetics, or Control and Communication in the Animal and the Machine,* Wiener wrote in his letter: "To provide scientific information is not a necessarily innocent act, and may entail the gravest consequences. . . . The interchange of ideas which is one of the great traditions of science must of course receive certain limitations when the scientist becomes an arbiter of life and death."[14] At the moment the bomb dropped, Wiener notes, scientists acquired the power formerly reserved for gods and kings; that is, the power to arbitrate life and death. Scientists, he argued, would need to reevaluate their methods and practices.

Research could no longer continue, wrote Wiener, in the idealistic but amorphous spirit of inquiry. Scientists would need to consider long-range ethics of particular technical developments, and limit access to such developments when they become too dangerous. Having developed a new science of self-regulating mechanisms, Wiener wrote, "we can only hand it over into the world that exists about us, and this is the world of Belsen and Hiroshima"; moreover, there is only slim hope that the benefits of cybernetics might "outweigh the incidental contribution we are making to the concentration of power (which is always concentrated, by its very conditions of existence, in the hands of the most unscrupulous)."[15] Wiener's caution, given voice shortly after the war and addressing the war's calamitous end, initiates several decades of serious public discourse concerning technological design. Yet Wiener's warnings were never followed as closely as his theories, or else the history of computation would be coterminous with a history of computational ethics. Alice Hilton's 1963 entreaty for a continued "awareness of the social dangers in the wake of our new technology" is therefore a direct effort to reopen Wiener's technological ethics in the face of its inevitable closure. Otherwise, very few traces of Wiener's admonition can be perceived in years that follow.

Popular alertness to the fact of computation does soon proliferate, however, along with hopes and fears about what effects teletechnology would wreak on the world of social relations. Captured in the phrase *global village,* the most influential of these ideas was adapted from a phrase by Wyndham Lewis. It takes on its contemporary sense in *Explorations in Communication,* a collection of cultural criticism and literary writing edited by Marshall McLuhan with Edmund Carpenter in 1960, where the editors write:

> Postliterate man's electronic media contract the world to a village
> or tribe where everything happens to everyone at the same time:
> everyone knows about, and therefore participates in, everything that

is happening the minute it happens. Television gives this quality of simultaneity to events in the global village. This simultaneous sharing of experiences as in a village or tribe creates a village or tribal outlook, and puts a premium on togetherness.[16]

Global village, a phrase now still much in circulation, thus combines an ideal of communication technology with an ideal of globally integrated life. Compressing both ideals into a historical argument and an ontological platitude, the global village is a purportedly unprecedented condition of existence in a new era. For McLuhan and Carpenter, it is accomplished fact: the new truth that diffuse locales and disparate chronotopes have come together in proximity and simultaneity.

In his own often-contradictory work, McLuhan goes back and forth on whether this technically enabled togetherness is actually a good thing, sometimes giving his critics sufficient reason to deride him for optimism, and sometimes claiming that the global village was only ever a dire state of affairs, protesting in 1967: "It never occurred to me that uniformity and tranquility were the properties of the global village. It has more spite and envy."[17] But whether the global village had coalesced through neighborly love, neighborly envy, or some combination of the two, there remains in McLuhan's work no question but that it had, in fact, coalesced. He clarifies: "I don't *approve* of the global village. I say we live in it."[18] Whether feared or celebrated, the global village has too often been accepted as common sense in political thinking and media study, as have McLuhan's dicta that technology is an extension of man and that the medium is the message.

But these dicta have not always been accepted. Kenneth Burke, for example, cast a critical eye on McLuhan's 1964 book *Understanding Media*, imagining how it might be rewritten as a theory less of sensible message than of sensory massage (a rewriting that McLuhan would in fact later embrace in his 1967 collaboration with Quentin Fiore, *The Medium Is the Massage*). In a review, Burke insists: "I'll gladly read it. Indeed, I can even glimpse some ribald fun here, based on lewd conceits about a man's extension."[19] Growing more serious, Burke then notes that McLuhan's whole project is made possible only by an increasing flexibility in its key terms, particularly those of information and communication: "If you give someone a hard blow on the head, this 'happening' can now be classed as a kind of 'information' that is physically 'communicated' to the nervous centers of the victim's brain," which is the kind of terminological muddiness that erases any "difference between an electric light and a comic book, or between a chemical and the 'iconic' image on a television screen" (413). Terms with formerly concrete and settled meanings became far more variable and contested after McLuhan. Or more precisely, for Burke, a theoretical nomenclature that had been subject to regimes of expertise—particularly in technical fields like electrical engineering and broadcast communications—was unleashed in ways that were highly inconsistent, yet readily adapted by nonexperts in politics and culture. The language of communication and information ceased to be chiefly denotative, and became increasingly connotative and ideological.

As Burke criticized McLuhan's shifting and appropriative lexicon, Raymond Williams aimed with equal energy at McLuhan's idea that the world had shrunk by telecommunicative means. In 1973, Williams insisted that the increasingly technological world is "not to be understood by rhetorical analogies like the 'global village.' Nothing could be less like the experience of any kind of village or settled active community. For in its main uses it is a form of unevenly shared consciousness of persistently external events."[20] The global village, to Williams, is an alibi. It provides cover for the varied developments in technology and economics and culture that make diverse peoples and diverse world-historical phenomena appear similar. Nothing has "contracted," power relations are "uneven," very little is "shared," and very few are "together." But then, if the world is not like a village, what is it like? Whether the metaphor of a global village is optimistic or pessimistic, it makes no admission of the historical violences—Wiener's world of Belsen and Hiroshima—that new technology had also produced. Whether proximity is a good thing or a bad thing, it assumes that race and class difference can be so easily reduced or eliminated. What kind of a world, it follows, might instead avow its differences and admit its violences?

In spite of objections by such public intellectuals as Williams and Burke, the village metaphor loses little of its near-universal appeal. Of the many who came after Wiener, writing contemporaneously with McLuhan and Carpenter, Buckminster Fuller is perhaps the best known. Just a dozen years after Wiener's warnings, Fuller attempted to blunt their effect, claiming that it is just such warnings that have limited the total, technologically enabled transformation of planetary life. He complained: "Our scientists are worrying about the exclusively negative and possibly lethal uses of their various special discoveries. At the same time we find society unable to translate the scientific discoveries into realistic magnitudes of comprehensive commonwealth advantage."[21] Simplistically, it might be said that Fuller expresses a technophilic rejoinder to Wiener's technophobic caution. Yet that would be to accept an unlikely proposition about Wiener, who was after all the principal developer of cybernetics. The difference is not between technophobia and technophilia, therefore, but instead between a thinker who has felt the weight of recent history (Wiener) and another thinker (Fuller, following McLuhan) who would set history aside in favor of a newly computational "universe capable of doing realistically unlimited work, ergo of producing realistically unlimited wealth."[22] For Fuller, only technophobic hang-ups or a lack of education can forestall the dawn of a "one-town world" in which "you will be able to go in the morning to any part of the earth by public conveyance, do your day's work, and reach home again in the evening, and . . . you will not have been out of town."[23]

Fuller's "one-town world" has very little to do Wiener's "world of Belsen and Hiroshima." Setting aside any barrier to technological imagination and invention, it shares much with another utopia of the time, the "Whole Earth" of Fuller's student Stewart Brand. Based in San Francisco, Brand founded the *Whole Earth Catalog* in 1968. Gathering together details on life in the new cybernetic reality, the catalog offered photographs of and instructions about everything that one might need to make art or machines, on or off the grid. It took its name from a button that Brand had designed and distributed in 1966, emblazoned with the question "Why Haven't We Seen a Photograph of the Whole Earth?" This question, and the phrase "whole earth" itself, condensed an idea much like that of the "global village" or the "one-town world": an idea of a self-contained sphere, not yet captured in a single photographic image, but soon to be thus captured, in which distance and difference had been superseded by a coherent planetary style of life. The *Whole Earth Catalog* was a guide to that life. Whereas Wiener cautioned scientists finding themselves suddenly to be "arbiters of life and death," Brand celebrated this new role: "We *are* as gods and might as well get used to it. . . . A realm of intimate, personal power is developing—power of the individual to conduct his own education, find his own inspiration, shape his own environment, and share his adventure with whoever is interested."[24] In place of Wiener's demand for an ethics of mutual obligation in the age of the camps and the bomb, Brand draws on Fuller and McLuhan to endorse a triumphant and unfettered individualism, and to publish its practical guidebook.

The futurisms of McLuhan and Carpenter, Fuller and Brand, are compelling stories. Such visions encompass the world as if it were one thing, a village that could be seen from space, a whole earth. They allow the Long Seventies to be seen as a period of consistent planetary feeling, whether that feeling be structured around teletechnological innovation, post-Woodstock defeat, corporate and technological ascendance, or coalition and mass mobilization. These varied futurisms are the principal midcentury manifestations of what Albert Szent-Györgyi called "cosmotechnics." Szent-Györgyi, an émigré Hungarian peace activist and a Nobel laureate in medical research, wrote in

1971 that the global imagination had made technological ethics all the more urgent. If the planet is now one thing, then the invention and application of machines must have implications that are newly planetary: "Cosmotechnics has created enormously powerful tools which enable man either to build a new world with undreamt of wealth and dignity or to altogether wipe himself off the surface of the earth. . . . It is the military, a relatively small group of people, which converts the results of science into instruments of murder and destruction."[25] Szent-Györgyi has something in common with his futurist contemporaries. He accepts that the world has indeed been wound up and shrunk down by new machines. But, like Hilton and Wiener, he differs from the futurists in foregrounding the problem of ethics, and in insisting on the role of technology in new forms of racism and war.

Whatever the similarities and differences, the stories told by all such public intellectuals exerted influence over technologically curious policymakers in media and cultural research. More than this, they informed counterparts in literary and activist writing, some of whom embraced the idea of a whole earth or global village, and some of whom followed Wiener's admonition that science should entail struggle in the world of Belsen and Hiroshima. In the present book, set in motion by Hilton's rereading of Wiener and Szent-Györgyi's critique of cosmotechnics, an ethico-political practice of activism and literature gathers together under the heading of cyberculture. Stubbornly refusing to be either utopian or dystopian, cyberculture likewise refuses to ignore the very real world in which machines are made and used.

Among forms of dismantling, cyberculture is a kind of missed option at the forking path of the technological imagination at midcentury. Instead of naming the new forms of speed or proximity that would be possible in the computational future, and instead of celebrating new forms of communication and data processing, cyberculture instead names the transformed conditions of human responsibility in a world of intelligent but nonhuman machines. In the terms of the word's coinage by Hilton in 1963, cyberculture offers an alternative to the global village and the one-town world, and an insistence on collective action in a world not only of Belsen and Hiroshima but also of ongoing struggles toward decolonization, sexual and gender autonomy, and racial justice. For Hilton, cyberculture is a challenge to those who would rather live without war, without racism, and without exploitative labor. It is, she argues, "our chance—'our' meaning all of humanity, not only a few who are exceptionally fortunate—to . . . do our human work, to live human lives, to devote ourselves . . . to poetry and politics."[26]

Luddism

Although it may appear opposed to any action taken against machines, Long-Seventies cyberculture has much in common with Long-Seventies Luddism. The former aims to make the best use of available technology, setting aside exploitative machines for machines that will reduce exploitation. The latter scans the landscape of new machines and aims to destroy the most exploitative among them, and from there to cultivate other kinds of growth. Cyberculture is more optimistic than Luddism, certainly, but the two terms differ more in their sequence (what shall we do first: examine or dismantle?) than in degree or kind. Both pursue, sometimes in tandem, a burgeoning of Hilton's pairing "poetry and politics." As Mitchell Goodman wrote in 1971: "It is time for a revival of Luddism—machine-smashing for the sake of survival."[27] What Goodman meant was not what has been meant by the word *Luddism* in recent years. Luddism is not the destruction of all machines. And neither is it the hatred of machines as such. Like cyberculture, it is another word for dismantling. Luddism is the performative breaking of machines that limit species expression and impede planetary survival.

Where had this "revival of Luddism" even come from? After all, the Luddites were a brief and historically specific political movement. Why should anyone care about the collective sabotage of steam-powered looms in the English Midlands? Beyond the evident similarity of material concerns between the second decade of the nineteenth century and the seventh decade of the twentieth century (as the war machine could be said to threaten life in the sixties in the way that the power loom had threatened British employment a century and a half before), interest in the Luddites seems to stem from the work of Eric Hobsbawm and E.P. Thompson, activist-historians of the postwar Left. Hobsbawm and Thompson, both Marxists, had advocated for a new and more favorable story to be told about the Luddites, in spite of Karl Marx's own opposition to their tactics. For Marx, machine-smashing, as the willful destruction of fixed capital, served as "a pretext for the most violent and reactionary measures" from the anti-Jacobin government, such that it was good when "the workers learnt to distinguish between machinery and its employment by capital, and therefore to transform their attacks from the material instruments of production to the form of society which utilizes those instruments."[28] While admitting Marx's point, Hobsbawm and Thompson supplemented his account, seeing no need to choose between an opposition to the instruments of exploitation and an opposition to the exploitative form of society. In a way that short-circuits much subsequent defense of technology,

Hobsbawm and Thompson find value in any act that would destroy both the machines and their legitimating institutions all at once.

Arguing in 1952 that for working people, "the basis of power lay in machine wrecking, rioting, and the destruction of property in general (or, in modern terms, sabotage and direct action),"[29] Hobsbawm concluded that "collective bargaining by riot was at least as effective as any other means of bringing trade union pressure, and probably more effective than any other means available before the era of national trade unions."[30] A decade later, in *The Making of the English Working Class*, Thompson concurred that Luddism exhibited "discipline and self-restraint of a high order" such that it proved "a manifestation of a working-class culture of greater independence and complexity than any known to the 18th century."[31] To say that Luddism was more successful than people remember, or that it interfered with the mechanisms of power, is not exactly to say that it is the revolution we need, either in the Long Seventies or today. But where the communities had been devastated by the power loom (as much as they would later be drained by a war machine), sabotage and "collective bargaining by riot" were useful tactics in the critical rejoinder to a culture of technology. For all that the Luddites failed to halt the advance of industrial capitalism, in Hobsbawm's words, "on a smaller scale . . . it was by no means the hopelessly ineffective weapon that it has been made out to be."[32]

When dismantling goes mainstream in the Long Seventies, first as cyberculture and then as Luddism, it is directed not at technology or science in any general way, and neither does it overlap fully with the newly emergent philosophical field, the philosophy of technology. For many writers and thinkers committed to social justice, technology is a signifier for any obstruction to group action, as well as an umbrella term for concrete tools of human auto-extinction. Luddism, in this respect, is a figurative response to capitalism and individualism, and a literal response to the self-destructive impulses of the species. In a 1976 article from *Fifth Estate* entitled "Who Killed Ned Ludd?," anarchists John and Paula Zerzan argued that Luddism would provide a way out of a deadlock in which unionism and fascism had become the twin horizons of the industrial and technological status quo. In a world where some with power have committed to "espouse unions as all that 'untutored' workers can have" and the rest with power have discovered how "fascism can successfully appeal to workers as the removal of inhibitions,"[33] there is no one left over to endorse "the unceasing universal Luddite contest over control of the productive processes."[34] Although unionists and fascists opposed one another, the Zerzans argued, both groups agreed to tolerate rather than destroy the tools of their exploitation. While unionists kept trying to negotiate an asymmetrical relationship with the owners of fixed capital, in other words, fascists had again begun calling for workers to embrace their own subjugation as if it were freedom. Luddism promised a third way. Aimed neither at destroying every machine nor at discarding the very idea of technology, Luddism aimed instead to break the array of particular technologies that get in the way of emancipation and survival. It offered a preamble to a collective analysis and invention that must follow.

In this book, suitably then, Luddism is two things: it is a destructive practice of machine-breaking, and it is a generative practice of ideology critique. As ideology critique, Luddism is the historical effort to mobilize literary and activist words against the hegemonic cultural formations—technocentrism and cosmotechnics, militarism and racism, gender and class exploitation—that rely on what Christopher Isherwood called "sickly idolatry" of computers.[35] As machine-breaking, dismantling names the extraliterary effort, as taught by certain texts, to break apart real machines, or to replace inappropriate machines with appropriate ones. And it is this project, the breaking of retrogressive notions of technology coupled with the breaking of retrogressive technologies, that undergoes a period of vital activity during the Long Seventies in the poems, fictions, and activist speech of what was then called cyberculture.

Cultivating Good Sense

In an ironic poem of the 1970s, entitled "Routine for a Stand-Up Comedian," Kenneth Burke writes:

> The guy said, "If machinery
> makes you so happy
> go buy yourself
> a Happiness Machine."
> Then he realized:
> They were trying to do
> exactly that.

> Patients are for hospitals
> authors are for publishers
> the law is for the lawyers.[36]

What motivates technological design and use, for Burke, is a desire for technology itself, rather than for human well-being. When he uses the word *machinery*, Burke means not only the literal mechanisms of science and industry, but also the institutions and devices of an increasingly technocratic society. Everything runs backward in such a society, for Burke, with the law serving the lawyers, rather than anyone it ostensibly protects; with medicine serving those who are paid for, rather than anyone it might heal; and with literature serving those for whom it is a widget to be made and sold, rather than anyone it might provide a political or spiritual benefit. The only people made happy by Burke's Happiness Machine are those who have been enriched by it. The Happiness Machine is not only the collection of particular industries that benefit from the enthusiasm for new technology; it is also a synecdoche for the political economy to which it is increasingly central. The Happiness Machine is both the tool and the sum total of technocratic capitalism. It is both the tech sector and the technocentric society. And until it is dismantled, the Happiness Machine will improve the lives only of those who own the means of production. The rest of us must look elsewhere for healing, protection, and enlightenment. The rest of us, along with Burke and Lorde and Hilton, must puzzle over a formula: claiming machines for the HUMAN BEING, not human beings for THE MACHINE, or asking which tools will dismantle the master's house.

In this book, the words *technology*, *tools*, *media*, *machinery*, and *machines* tend to overlap in meaning, becoming indistinct, and there is a good reason for this. Literary and activist texts rarely bother with the terminological distinctions of philosophers. Because what follows is an engagement with literature and activism, more often than with philosophy, I follow the lead of texts I examine. I take frequent advantage of the terminological dissensus, allowing these terms to drift in and out of their familiar definitions. I talk about technology as a way of getting at something else, something that is media-implicit not media-explicit, and not the essence of technology. I sideline or criticize claims to the revolutionary or apocalyptic capacities of electronic and communicative media—that is, of the technologies themselves—because it is more important to foreground the political and aesthetic and social discourses that circulate through a world that includes technology, or that might be reimagined as itself a technology, but that will not be reduced to a technological process.

Technologies, in any case, are already social, whether one is talking about things that are obviously technological, like labor, war, representation, communication, data processing and storage, energy, and transportation; or else about things that are less obviously technological (those considered technological only by philosophers and critical theorists) like artworks, institutions of thought, political procedures, clichés, systems of cultural difference, and the very idea of a functionally unitary world. Practices of dismantling, of cyberculture or Luddism, may be directed at any of these. Such practices thus dovetail with work by left thinkers like Zielinski or Lorde, Langdon Winner or Donna Haraway, whose voices can be heard echoing throughout this book. What technological theory of literature might survive in the shadow of the demand to dismantle the master's house? Concomitantly, what literary theory of technology might facilitate solidarity work across social movements, while also seeking to examine and then discard the master's tools? This double question, in pursuit both of hopeful and of skeptical visions of literature, is precisely the path of literary politics. It is a path laid in the writing of literature and in its extension to the essayistic traditions of radical political culture, and occasionally in the transforming (in both transitive and intransitive senses of that word) practices of literary studies in the academy. It concludes, following David Harvey, that "there is no such thing as a good and emancipatory technology that cannot be co-opted and perverted into a power of capital"; and it finds symmetrically that no reified notion of technology can serve as a bad-enough villain for new revolutionary stories to be told in this present.[37]

In most of its uses, the language of technology tends to stand in for other things that are not, or not obviously, technological. It is a language of metaphors, casting present problems in terms of imaginable solutions. Metaphors are not merely different from concepts. In their multiplicity and indeterminacy, metaphors can also undermine concepts. When theory is undertaken as a way "to produce viable action," philosopher Kristie Dotson writes, "literature, poetry, autobiography would count as viable sources for philosophical engagement. . . . That is to say, valuing multiple forms of disciplinary validation acts like a check against the universalizability of definitions of philosophy and their resulting justifying norms . . . that are falsely taken to be commonly held and univocally relevant."[38] Technology exerts a gravitational pull on contemporary discourse in this digital age. But literary metaphors and activist polemic may yet resist this pull, by refusing any universalizing take on technology, abandoning any refined concept that claims, in Dotson's terms, to be "commonly held and univocally relevant." This said, there is no point in being coy about definitions. The main target of my analysis is the central cultural roles that are granted to what are usually called "technology" and "communication."

Such a theory, whereby poetic practices interfere with technological practices, introduces a perspective that gives more credit to the nonphilosophical ideas of technology than to the ideas of technology that are developed in the subfields of critical theory. It can be difficult to bring such a theory, stripped of the familiar philosophical verities, back into the domain of cultural theory. In the U.S., most such theory (especially after the

International Colloquium on Critical Languages and the Sciences of Man at Johns Hopkins in fall of 1966) is built not on the antagonism of *poiēsis* to *technē*, but instead on the inseparability of these terms. Heidegger wrote famously of the knot they tie: "*Technē* is the name not only for the activities and skills of the craftsman, but also for the arts of the mind and the fine arts. *Technē* belongs to bringing-forth, to *poiēsis*; it is something poietic"; and his account is etched deep in the phenomenological and deconstructive accounts of technology in later years, nearly obstructing any chance of another account.[39] If technics belongs to poetics, and actually "is something poietic," then how will procedures associated with *poiēsis* ever sabotage any procedures associated with *technē*? How will art or poetry help dismantle the worst outcomes of invention or innovation?

Even in the archive of largely French theory that absorbs Heidegger's indistinction and then exports it to the United States, however, there is criticism that remains committed to wielding radical poetic thinking against reactionary technological thinking. Central to such criticism is Jean-François Lyotard, whose report in 1979 to Quebec's Conseil des Universités has become canonical as *The Postmodern Condition: A Report on Knowledge*. Usually cited for having identified the end of "grand narratives" (an observation undercut by the fact that it is itself such a narrative), Lyotard's book is now most valuable as a critique of science and technology at the dawn of a newly computational present. Before the contemporary era, writes Lyotard in that book, it made sense to see technology as the set of tools belonging to the practice of science. Tools for measurement or for the extension of human capacities, technical devices seemed capable of facilitating human knowledge forever. Thus in bygone days, following a "principle of optimal performance," technology could be defined simply "as prosthetic aids for the human organs or as physiological systems whose function it is to receive data."[40] However, writes Lyotard, this all changes in the industrial revolution with the introduction of new systems whose only purpose is to maximize wealth over time. Technology was no longer just the set of tools that facilitated science. Instead, industrial technology became the driver for institutions and methods of science, making it so that "science becomes a force of production" and "a moment in the circulation of capital."[41] Under the sign of capitalism, science thereby became more and more of an alibi for technocratic schemes that led through efficiency toward wealth, as Lyotard concludes: "Thus the growth of power, and its self-legitimation, are now taking the route of data storage and accessibility, and the operativity of information. The relationship between science and technology is reversed."[42] Technology no longer serves science, that is, so much as science serves technology.

A tendency that began in factories, with printing presses and mechanical looms, thus reaches its apogee in computation a century and a half later. This is the tendency that triumphs. Having won a contest against anybody who would conduct research for other reasons, industrialists then pass the spoils down to their computationalist heirs. Capitalism gets to do what capitalism has always done, which is to put knowledge at the service of technology rather than the other way around. Yet in the face of such crushing loss, writes Lyotard, anticapitalists and noncapitalists do manage to develop contrary research practices: for while research as "a matter of curiosity and artistic innovation" had thrived from the sixteenth to the end of the eighteenth century, "it can be maintained that even today 'wildcat' activities of technical invention, sometimes related to *bricolage*, still go on outside" of the domains of operativity and accessibility.[43] A period of wildcat strikes thus also produces wildcat technical practices. Technology as artistic innovation, combined with the art and theory that undermine power, joins a centuries-old resistance to the rules of efficiency and productivity. Following Heidegger, it may be philosophically true that technics are poetic at some register, or that poetry is technical. But following Lyotard, it is more important that technocratic priorities can still be dismantled by "wildcat" practices of justice or art. It is for this latter reason—the dismantling of what is usually called technology with the devices that are usually called poetic—that these terms can remain opposed.

Even before Lyotard, Theodor Adorno too (despite the ample differences from Lyotard that Lyotard explores most famously in "Adorno as the Devil") produces an aligned critique of technology. Taking aim at the ideological formations that develop under an advancing technological capitalism, Adorno considers ways to oppose an increasingly machinic modernity. These tendencies consolidate in a mythmaking procedure that he calls the "technological veil." Whatever the benefits of particular machines, Adorno argues in 1969, "there is something exaggerated, irrational, pathogenic in the present-day relationship to technology. . . . People are inclined to take technology to be the thing itself, as an end in itself, a force of its own, and they forget that it is an extension of human dexterity."[44] In truth technology is not veiled. Rather technology is itself a veil, concealing modes of production, presenting itself as ends rather than means, and contributing to the collective alienation from real material relations. Technology, for Adorno, is not merely pathological, as if its veil were a disease of perception; rather, it is pathogenic, generating all manner of secondary and tertiary ailments. The costs of this pathogenic capacity are high: "This trend goes hand in hand with that of the entire civilization. To struggle against it means as much as

to stand against the world spirit (*Weltgeist*); but with this I am only repeating . . . the darkest aspect of an education opposed to Auschwitz."[45] In the
dedication of the world spirit to belief in the technological veil, the species
has entered into vile contracts with its machines. Humanity can thus accept
even genocide, if genocide presents itself as the full utilization of available
machines. To struggle against this acceptance, to buck humanity's eager sub-
ordination to technology, is therefore to educate oneself against Auschwitz.

Machines are not themselves to blame, Adorno cautions in 1968. Or
rather they are not to blame simply because they are machines:

> It is not technology that is the catastrophe but its imbrication with the
> social relations that embrace it. We should merely remind ourselves
> that it is the concern for profit and domination that has canalized tech-
> nological development: on occasion it coincides in a disastrous way
> with the need to exercise control. Not for nothing has the invention
> of weapons of destruction become the new prototype of technology.
> And, by contrast, those technologies that turn their backs on domina-
> tion, centralism, and violence against nature, and that would doubtless
> help to heal much of what is damaged literally and figuratively by the
> technology we have, are allowed to wither away.[46]

Long Seventies literary politics resides here, within Adorno's dual problematic. On one hand, in cybercultural language, it may be observed that technology is not to blame for human exploitation except when, in intercon-
nected fashion, it is applied to exploitive ends. On the other hand, Luddites
can insist that most technologies are applied in exactly this way, to the extent
that military technologies are the paradigmatic ones and any dissenting or
ameliorative technology tends to get abandoned or suppressed. Prose and
poetry are not ontologically inclined—or at least not any more inclined than
other zones of the culture industry—to redirect the canalization of technological development. But neither is poetics just a facet or effect of technics.
If written words can transcend their own mode of production, even just long
enough to "heal" much of what is "damaged," then they deserve to be read
and understood.

Literary and activist accounts of technology during the Long Seventies,
although they are rarely in conversation with Lyotard and Adorno, often
resonate with their critiques. They put pressure on the smooth functioning
of institutions and devices, and press back at mythmaking procedures, by
means that are both literal and figurative. In these accounts, technology is
the driver and material expression of scientific method and rational inquiry.
It is also the mechanism of industrial and postindustrial exploitation and

production, with no fully abstract definition removed from its instantiation by real machines. Technology is therefore defined not only as applied science but also as machinery—as what Marx calls a "colossal assembly of instruments."[47] In the shape of such an assembly, technology provides the "appropriate form of the use value of fixed capital," by which Marx means that it names the lossless value emitted by the industrialist's almost-permanent equipment.[48] In this literal sense, technology serves its owner, while providing a durable store of value through which its owner can retain a disequilibrium of power over the workers who actually operate and maintain the same machines. As a result, technology is what those operators might smash or seize when at last they undertake revolutionary activity.

In what follows, technology also has a figurative meaning. Technology poses a problem of knowledge for a society that has backed itself into an ethical and political corner. At the same time as technology is taken to have done damage to the human species and its environment, it is equally accepted that only technology will repair that damage. As Burke put it satirically in 1974, technology is at the center of a commonly held and naive hope: "If but technology continues to proliferate as it is now doing, things can end, not in a reactionary rejection of technology . . . but in a super-technology that can rise out of the very decay it is producing."[49] There is in the Long Seventies a growing sense that—workers having been exploited or killed at the switch or on the line; environments having been devastated by industrial waste; communities shattered by architecturally miraculous highway bridges; sexed and gendered bodies controlled or indentured by a heteromasculinist state; cities and populations leveled by bombs; racial and ethnic groups decimated by high-efficiency genocide camps—technology has done its job too well. Its horrors result not from poor technological design but instead from its designated and ostensibly appropriate use. Its misapplications are not misapplications at all. It is the figure for humanity's systematic self-elimination.

With less irony than Burke, but giving perhaps more cause for optimism, Marxist theorist Moishe Postone addresses a similarly figurative problem of ethics and politics in 1978. Postone identifies

> two antinomic socially critical positions which are immanent to the capital-determined social formation: (1) the attempt to "overcome" alienated labor and the alienation of people from nature by rejecting industrial technology per se, in the hope of a historically impossible return to a preindustrial society, and (2) the attempt to strive for a just mode of distribution of the great mass of goods and services produced, while accepting the linear continuation of capital-determined technology in its manifest form.[50]

Postone's "critical positions" emerge from a common acceptance of the existing conditions of technological design and application, but each is twinned to one of the revolutionary practices, Luddism and cyberculture, that concern this book. The "rejection of industrial technology per se" is nothing but a failed version of Luddism. Where the former would return to the past by breaking all machines, the latter would, selectively and strategically, shut down reactionary machines in a confrontation with the present. Likewise, the effort to come up with technological solutions to resource management, but without breaking with capitalist modes of production, is a failed version of cyberculture. Where the former would seek other inventions by which to redistribute the material products of a reformed capitalism, the latter would aim for other ways of inventing and distributing aside from capitalism.

Postone sees these compromised critical positions as immanent to capitalism, and in need of "overcoming and rejection."[51] Luddism and cyberculture are names for this overcoming. Were such practices able to emerge through a sublation of the capitalist formations that make them possible, they would not only change the conditions of labor in a technologized culture—they would change technology itself: "The course of capitalism drives technological development . . . , whose concrete form remains an instrument of domination—yet whose growing potential allows for a transformation of society . . . such that *not only the goal of machine production but the machines themselves will be different.*"[52] Luddism and cyberculture, in their nonnaive and nonvulgar forms, can emerge only from the capitalist development of the instruments of domination that they oppose. But once they have emerged—that is, once patterns of sabotage can proceed without targeting technology per se, and once resource distribution can proceed without accepting their original maldistribution by capital—the figurative puzzle of technology will be irrevocably changed, perhaps even solved.

Keeping in mind this revolutionary imagination of the Long Seventies, a new theory of technology might be practiced: instead of a progress narrative of technological evolution and the achievement of global unity by telecommunicative means, and instead of a rigid lexicon for the study of technology and society, we may yet begin to understand what connects literal and figurative technologies to the cultural differences, complex forms, flexible vocabularies, critical analyses, and revolutionary communitarian possibilities of our present. Those years were often hard for activists and artists on the Left. Gil Scott-Heron once explained the meaning behind one of his best-known songs, describing a period in which "the forces that in the '60s were drawing people together were assassinated and repressed [while] a climate was brought in which was receptive to law and order and phone-bugging. . . . Ever since then it's been Winter in America."[53] Scott-Heron is optimistic about the potentiality of struggle, concluding that "you go through winter to get to a new springtime." Still, it is worth acknowledging that scholars of U.S. culture do not often associate the 1970s with any such widespread ambitious rethinking of literature, technology, and the politics of solidarity. Instead, too often, they treat it as a kind of dead zone, or as a period of decline before the defeat of New Left politics. With few notable events aside from the tumultuous finales to the Vietnam War and the Nixon presidency, the decade is treated as a fermata, or pause, between the unconstrained sex of the sixties and the unconstrained capital of the eighties.

The poet-critic Stephen Paul Miller writes of this problem in his own address to the decade, *The Seventies Now: Culture as Surveillance*: "When I told a historian that I was writing a book about the seventies, she asked me how I could do it. 'Wasn't Watergate the only thing that happened during the Seventies,' she asked? . . . A conventional history of the seventies might have difficulty accounting for the vast sea change that the nation underwent during the decade. In typically positivist terms, it is difficult to account for how the decade of Watergate set the stage for the Reagan administration."[54] Miller's solution to this "difficulty" is to focus inquiry on a single transformation belonging to the decade: the shift from a dominant culture characterized by surveillance-by-the-state to a dominant culture characterized by surveillance-by-the-self. Like Miller, I am not a historian; and like Miller, I focus on a single transformation. Yet whereas Miller focuses on dominant culture, I focus instead on insurgent culture: the "politics and poetry" of Alice Mary Hilton, say, or the not-the-master's tools of Audre Lorde. Whereas Miller accepts the centrality of a mainstream political paradigm (Nixonian surveillance), I set aside electoral politics entirely. And whereas Miller focuses on the popular introjection of the surveillant gaze of the state, transforming fear into anxiety, I focus instead on political introspection within literary tendencies, like cyberculture and Luddism, that oppose the state in an effort to turn hope into strategy.

When it said that the seventies are themselves a period marked by failures of the Left, following the near victories of the previous decade, critics generally point to May 1968 in Paris as the most important convergence of those near victories. But the *Mai 68* protests were anticapitalist and anticolonial in their focus, whereas protests in the United States coalesced just as forcefully around feminism and racial justice. If, as Kristin Ross puts it, "the clear ideological targets [of the Paris uprising] were three: capitalism, American imperialism, and Gaullism," then *Mai 68* can offer only a partial model for radical action in the United States.[55] In the U.S., this Parisian time-space provided both an inspiration and a cautionary tale. But so did the Detroit riots a year earlier and the Stonewall uprising a year later. So did the election of a socialist government in Chile the year after that, and the U.S.-aided murder of that government in 1973. So did the persecution of Angela Y. Davis, and the state assassinations of Fred Hampton, Pablo Neruda, and Steve Biko. So did the ongoing wars in Israel-occupied Palestine all those years, from the Six-Day War and War of Attrition to the Yom Kippur War. So did Woodstock, it is true, but then so did Altamont. So did My Lai. Perhaps the most fundamental change before and after May, in both France and the United States, was an increased alertness to the mutability and resilience of state power. So it is that Ross finds after May "a period of massive concern with public order and its breakdown, when the government's tangible fear of the population taking to the streets again had manifested in a dramatic increase of police presence everywhere."[56] For Ross, this concern and threat echo as primary concerns of philosophy and theory in France during the years afterward. For the present book, they echo equally through the technocritical literary and activist writing of the United States, yet emanate as much from Detroit and Stonewall (and Gaza, Altamont, Santiago, Chicago, My Lai, Palo Alto, and elsewhere) as from Paris.

As a record of this concern and threat, of a textual experience of order and its breakdown, of protest with and against technology, *Dismantlings* is one part archival rescue mission and one part critical theory of literature and technology. It is not really a history, although at times it tends toward genealogy. Instead it is a textual analysis of the past and an analysis of past texts. Motivating its inquiry is an impulse less like Miller's and more like the one that motivates Elizabeth Freeman's 2010 book *Time Binds: Queer Temporalities, Queer Histories*. Like Freeman, I pick through "cultural debris" of "incomplete, partial, or otherwise failed transformations of the social field."[57] As her book is, this one is "organized not around the great wars of the twentieth century and beyond, but around a series of failed revolutions in the 1960s and 1970s—political programs not only as yet incompletely realized but also impossible to realize in their original mode—that nevertheless provide pleasure as well as pain."[58] Like the revolutions that occupy Freeman, and indeed involving them at points, the revolutions of dismantling are efforts to think and act against a prevailing common sense.

Because today's common sense is so different from that of the Long Seventies, especially when it comes to the social and cultural formations that surround new technology, there is no point in trying to defibrillate the old, dead thing. If no political overturning followed the literary politics of cyberculture and Luddism in their own moment, then certainly none will follow them now. But there is pleasure and pain, to borrow Freeman's words, in retrieving coherent and righteous thinking from the cultural debris. There is filiation to be felt with any prior thinking that would substitute communal technologies for exploitative ones, or consign exploitative technologies to the junk heap. Even as a revolution fails, its failure fuels common feeling without which subsequent revolutions cannot succeed. No procedure of art or activism can achieve human ends by inhumane means, but thankfully that is not the only option. Dismantling is a way to understand new tools without embracing or rejecting them on principle. It is a way to understand the world as something other than a communications network, and a way to undermine the ideas about technology that underpin racism, settler colonialism, class exploitation, and the gendered division of labor.

The promise of instantaneous long-distance message transmission—and of its cosmotechnic counterparts like the "global village," "one-town world," and "whole earth"—took root in the Long Seventies, linking liberal social ideals to developing technologies of communicative speed and transparency. What, I ask, would it mean to break this promise, on behalf of marginalized peoples and marginal forms of life? To forget the image of a singular planet, instead to track perspectives on technology through what Raymond Williams called the "unevenly shared consciousness of persistently external events"? To break technology, or to break with technology, in the moment when breaking becomes necessary? To avow the world of Belsen and Hiroshima, while insisting that it also be a world of politics and poetry?

A Counterlexicon

This book's argument is its archive. It is an archive of ideas and trajectories, variations on the knowledge of the present and techniques for its study, words wound around one another like threads in a rope, or arrayed like a constellation:

Luddism. A metaphor that is not only a metaphor, *Luddism* threads through the Long Seventies in the work of poets, activists, and thinkers, each of whom applies literature to the task of dismantling the technocentric world. For example, whereas Édouard Glissant may offer an optimistic promise of literature's power to break systems, writing that poetry can "thicken" the "machine that the world is," Audre Lorde is later more skeptical, opening the possibility that even literature may be one among the "master's tools" that are inapposite to the task of dismantling. Joanna Russ is more skeptical still, in her insistence that scholars and science-fiction writers should "give up talking about technology," as such, and W.S. Merwin imagines an intelligent machine that is fated to be relinquished. Crucially, such literary and theoretical practices do not oppose technology as such, but instead oppose large-scale forms of exploitation by dismantling the machines at their disposal. To such critiques, Langdon Winner gives the name of a potential philosophical method: "epistemological Luddism." Epistemological Luddism is that specific form of Luddism that provides a critical defense against late-twentieth-century technological politics and a dedramatization of the false choice for or against technology.

communion. A planetary practice of collective self-identification, communion takes shape in opposition to teletechnological ideals of global togetherness. Opposed to the cosmotechnics of Spaceship Earth, a metaphor devised separately by Buckminster Fuller and Adlai Stevenson, communion is better developed in science-fictional work of Samuel R. Delany and Ursula K. Le Guin. Spaceship Earth, along with aligned metaphors of technologically enabled proximity, is in part to blame for sustaining the fiction of a world fused by common cause, and for perpetuating the accepted language of techno-boosterism, even in left cultural critique. Communion, by contrast, sees little such common cause in the world, but instead sees the cosmotechnic globe as the contested ground for coalitional struggles for real coexistence.

cyberculture. The interconnection of transformations in cybernetics (automation), weaponry (the atom bomb), and human rights (antiblack racism), cyberculture was coined by Alice Mary Hilton and then extended conceptually and politically by the Ad Hoc Committee on the Triple Revolution. This committee, in 1964, mailed a letter to Lyndon Johnson, explaining that an expansion of industrial automation might result in an alleviation of racism and war. With a luminous gang of signatories—antiracist activists James Boggs and Bayard Rustin, SDS leaders Todd Gitlin and Tom Hayden, literary critics Irving Howe and Dwight MacDonald, public intellectuals Gunnar Myrdal and Linus Pauling, et al.—this letter is, on its face, a mere precursor to contemporary accelerationist movements. Yet in its more immediate effects, not as an endorsement of accelerated automation but instead as a critique of technologized violence, it provides a conceptual substance that permeates literary and activist writing of the Long Seventies. The theory of cyberculture reveals something beyond accelerated automation. In a world that is still defined largely by the application of advanced tools to sociopolitical problems, cyberculture reveals, the question to ask is not: Which tools ameliorate and which delimit the lives of humans and nonhumans? The question to ask after cyberculture is instead: Which technological practices can be reconciled with which practices of peace and community?

distortion. The science fiction writer Samuel R. Delany insists that transformative change takes shape neither in utopian nor in dystopian visions of the future, but rather in efforts toward significant distortion of the present. This attitude, which is also a theory and practice of literature, is one way to describe the inheritance of cyberculture in the works of writers and activists who employed speculative language to repurpose the thought of Hilton and the Ad Hoc Committee. These writers and activists, from Martin Luther King, Jr., and Shulamith Firestone to the poets James Laughlin and Thomas Merton and the science fiction writer Philip José Farmer, focused not on the machines that would unveil the myth of scarcity, but instead isolate the forms of human life and relation that would follow that act of unveiling. As a form of dismantling, distortion is the historical and theoretical technique by which readers learn to approach political documents as if they were science fiction. When considered as a vehicle of distortion, literature is measured for its potential to alter exploitative conditions, like those of war, patriarchy, and racism.

revolutionary suicide. Huey P. Newton's phrase for a form of political commitment that includes a willingness to endanger oneself. Not an advocacy for death, revolutionary suicide is the idea of a furious collective survival at all costs. As forms of dismantling, the theories and practices of revolutionary suicide demonstrate how bodies may strike not only against their machines, but also against themselves, if the alternative is to be made into a machine. Revolutionary suicide is the cybercultural self-elimination of one body in response to instrumentalization by another. To resist being made into a machine, to be used by someone else, is to ask: What kind of dismantling might be directed at oneself so as to end one's bondage? This question winds like a wire through literary experiments by Russ, Paul Metcalf, Thomas Pynchon, and Toni Morrison. Pynchon would later write an op-ed for the *New York Times*, entitled "Is It O.K. to Be a Luddite?" To that question, but several years before, these novelists answer yes. In a mode of revolutionary suicide, dismantling is a generative protest: not against technology but instead against the instrumentalization of human life through techniques of compulsory motherhood, black slavery, militarized science, and the binding constraints of humanist fiction. In characters' capacity to dismantle themselves, within and against the conventions of genre and story, revolutionary suicide is how texts rupture the stultifying categories of race and reproductive technology in defense of subjugable bodies.

liberation technology. A term for a situated theory of communitarian tool-use, developed in the activist and philosophical work of Seneca leader John C. Mohawk, also known as Sotsisowah. Discrete from other ways of thinking about machines and freedom (like the cyberlibertarianism to which Ayn Rand gave voice in the same years), liberation technology more closely resembles the techniques and tools of what Maria Mies and Carol DeChellis Hill, in very different registers, called a politics of subsistence. When politics of technology and survival are seen as largely local, there remain ways to flourish at a subsistence level on the outer edge of the technologized world. Liberation technology grants that there is both value and risk in having nothing, or almost nothing. It names the basic material and collective strategy that can facilitate a transformation of shared life even under extremely exploited conditions. It is a hybrid concept, coined by Mohawk in a speech before the UN in 1977, that synthesizes liberation theology with appropriate technology, and might lead to unexpected kinds of shared belief and action.

thanatopography. Naming a spatio-temporal configuration of the world that opposes cosmotechnics, the word itself originates in the fiction of Stanley Elkin. Not thanatography, which is the writing of a death, thanatopography is the drawing of a map of death. When new technologies respatialize the world, thanatopography teaches that they do so not because they construct a communicative network but instead because they build and distribute sites of machinic killing. In returning to Norbert Wiener's insistence on seeing the planet as a world of Belsen and

Hiroshima, thanatopography enables us to pare back the presumed connection between technology and humanity; and to expose something quite frightening underneath the network. A vision of the world that presumes no common similarities among people and peoples is a vertiginous vision that must see shared connections among extant technologies, not only telecommunication and computation but also war, racism, and dehumanizing labor. Communal responsibility, yes, mutual obligation, yes, these survive amid such technologies as ethical codes that negotiate difference rather than attempting to transcend it. But they also require a reckoning with very real legacies of twentieth-century machines. In place of the smooth-functioning global network, these texts offer a spatio-temporal figuration of mass death.

CHAPTER 1

Luddism

*one by one, but with growing frequency, [they] will begin to
lose their machines*

—W.S. Merwin, "The Remembering Machines of
Tomorrow" (1969)

In 1969, *The New Yorker* published an experi-
mental prose poem by W.S. Merwin on the topic of computation. In a series
of ruminative paragraphs called "The Remembering Machines of Tomorrow," Merwin describes a scene that was then fictional but seems now perfectly common. Humans delegate a human task, memory, to a nonhuman
machine and soon come to rely on that machine. In an unfamiliar coda to the
familiar narrative, the poem ends when the humans finally cease using their
machines. The piece does not narrate a refusal of information technology.
Instead it invites a gradual relinquishing of it: "Attached to every person like
a tiny galaxy will be the whole of his past—or what he takes to be the whole
of his past. His attachment to it will constitute the whole of his present—or
of what he takes to be the present. The neat, almost soundless instrument
will contain all of each man's hope, his innocence, his garden. Then one by
one, but with growing frequency, men will begin to lose their machines."[1]
Merwin's "instrument" has a capacity for prosthetic memory that gives way
to transformations in thought and spirit, because the remembering machine
is "attached to every person" such that it "constitutes the whole of his present." Yet the remembering machine will ultimately be lost, surrendered to
disuse. Its owners will learn to live without it, or else forget how impressed
they ever were with it.

Merwin's poetic image—the image of a fantastic tool that is finally abandoned—is a dedramatization of both the technological and the antitechnological impulse. This is Luddism of a sort. The machine vanishes without hammers and without malice, but it does nevertheless vanish. It is not that kind of Luddism that is associated, say, with the nostalgia for old tech like pay phones and arcade games; less still with what Dominic Pettman dismisses as that "nostalgic, romantic, neo-Luddite position (which assumes, naively, that the human somehow predated technics, before being contaminated by it)."[2] Rather, it is a gradual relinquishing of machines whose continued use would contravene ethical principles. To let go of certain technologies in order to let go of certain forms of dehumanizing activity—this is instead the activity that Audre Lorde calls *dismantling* and Langdon Winner calls *epistemological Luddism*. In much of Long Seventies cyberculture, as in Merwin's poem, this activity is accomplished by literary means, such that poetic tools are indeed mobilized to interrupt technological values.

Dismantling the Master's House

Perhaps the clearest account of dismantling-as-relinquishing is Audre Lorde's 1979 declaration that "the master's tools will never dismantle the master's house"—in her lecture of the same title at the University of Kansas in 1979, during the "Second Sex Conference," the first meeting of the National Women's Studies Association. Lorde's influential claim was that feminist practices of mobilization and analysis must account not only for gender difference but also for differences of class, sexuality, age, and race. By working and thinking on behalf of women yet simultaneously on behalf of people who are poor, queer, old, or black, Lorde imagined a movement that might possibly break the devastating political technologies that had partitioned those forms of identity from one another in the first place. The metaphor of tools is not incidental. Lorde stands at the near end of traditions in feminist and antiracist poetry and oratory, but she stands as well in the tradition of the antitechnological Luddites at the beginning of the nineteenth century, and in the tradition of subsequent revolutionaries for whom acts of sabotage can signal a total detachment from the reactionary norms of society and culture.[3]

In its central image—the search for correct tools to break the machines of the house of power—Lorde's paper hinges on two distinct claims about technology: one, not all tools are appropriate to all tasks; and two, some worldly objects (in this case, the master's house) cannot be reinvented but must be destroyed. Whether or not the proper tools may be found for dismantling the house, this much is clear: both master's house and master's tools must be abandoned.

The tools of a master, Lorde writes in her lecture, "may allow us temporarily to beat him at his own game, but they will never enable us to bring about genuine change."[4] Other tools will be required, other tools than the inherited ones, if solidarity is to be achieved. Call them machines or technologies or devices, tools are Lorde's metaphor for instruments of power as well as the opposition to power; for what gets dismantled as well as what dismantles; for institutional language as well as analytic thought; for the extant world as well as the tactics that might change the world. Dismantling, following Lorde, is a labor of picking apart social machines in order to understand them, and an insistence on smashing those antisocial machines that sustain unequal distributions of power. Lorde sees feminist texts and institutions as largely blind to forms of structured power aside from gender, even though most women's lives are as thoroughly structured by inegalitarian structures of "race, sexuality, class, and age."[5] Blinded in this way, Lorde claims, feminists fail to recognize "the interdependence of mutual (nondominant) differences . . . enables us to descend into the chaos of knowledge and return with true visions of our future."[6] The "master's house" is where the heteropatriarchy, classed and raced, thrives and reproduces itself. It is where women are treated, and treat one another, only as women. It is therefore, for Lorde, the very house that must next be destroyed.

Here is where the logic of technology, considered as the logic of tools and the regimentation of the terms of tool use, comes in. Thinking about social and political power as a kind of technology or machinery is a procedure many centuries old, so the technological aspect of the master's tools metaphor is not of itself a surprise.[7] What surprises is the sheer flexibility of technological metaphors, such that they might ground emancipatory claims like Lorde's, but also justify a more reactionary idea of machines in and of the world. Lorde asks: "What does it mean when the tools of a racist patriarchy are used to examine the fruits of that same patriarchy?" And then answers: "It means that only the most narrow parameters of change are possible."[8] To Lorde, tools are for examining and dismantling, more than for making, and their most urgent task is to examine and dismantle the epistemological impediments to solidarity. There are tools that can aid in examining and dismantling, and there are tools that are not adequate to that task. The inadequate tools are the "master's tools" that, because they are the "tools of a racist patriarchy," are suited only to the preservation of that patriarchy.

Meanwhile, what sets Lorde's speculation in motion is that there may be tools still more adequate to antiracist feminism—tools, in short, that emerge neither from racism nor from heteropatriarchy, nor from ageism, nor from capitalism. When Lorde describes feminist solidarity, she is modeling a kind of togetherness distinct from the contemporaneous metaphors of cosmotechnics. Whereas the global village (or the one-town world or the whole earth) comes together through a machine-enabled shrinkage of the world, Lorde's world comes together without shrinking, through the breaking of partitioning machines rather than the making of communicative ones. Lorde's aimed-for political movement can be accomplished only by people whose bodies are different from each other (aged, classed, sexed, as well as gendered) but whose purposes converge in opposition to anybody who would deny that difference. Her twinned metaphors, the master's tools and master's house, are not generally considered in light of technological transformation. But read as a form of dismantling, Lorde's metaphor remains a critique of cosmotechnics. Teletechnology and military technology have only yet produced illusory regimes of global connectivity and capital flow. Dismantling seeks precisely to dispel those illusions and help topple those regimes; and, for Lorde, it can be accomplished by poetic means.

Intimate Scrutiny

Charlotte Bunch precedes Lorde in metaphorizing tools for feminist action, in her well-known 1974 essay "The Reform Tool Kit." Distinguishing reform from reformism, Bunch argues that the former is very often subordinated and appropriated by the latter. Whereas reform is the transformation or demolition of existing institutions, she writes, reformism is only a reaction-formation: a backing away from revolutionary principles, and a strategic accommodation of violent institutions. For Bunch, the biggest challenge to feminism is that reformists have co-opted reform. The consequent need, therefore, must be "to develop a new kind of politics that could not be co-opted."[9] This is Bunch's version of Lorde's master's tools argument. Bunch and Lorde, it is often said, gesture only vaguely, suggesting rather than strategizing, toward the destruction of unnamed old systems with unnamed new tools. But it is only a misreading that can cast Bunch's reform-tool-kit or Lorde's not-the-master's-tools as a simple opposition to all forms of political dominance. Both instead issue specific calls to specific kinds of action. Just as Lorde's principal objective is to knock down the partitions that separate antiracist from antipatriarchal (as well as working class, antiageist, and anti-homophobic) struggles, Bunch's principal objective is to recover the feminist tactics that have gone unused because of their co-optation by mainstream liberal feminism. For Bunch, developing a tool kit of useful strategies means making an admission—that "we need a new social order based on equitable distribution of resources and access to them in the future"—and asking a set of guiding questions, including: "What kind of process does this involve?" and "What types of power must women have to make these changes?"[10] The tools in Bunch's tool kit are an egalitarian polemic and a pile of open-ended questions; not just solidarity in a vague sense, but specific weapons against "the slimy institutions we want to destroy."[11] Bunch's sense of inquiry, simultaneous with her sense of urgency, is what prefigures the question implied by Lorde's title: By what tools, not those of the master, is the master's house to be dismantled?

Lorde has her own answers, as is evident across the text of *Sister Outsider*, her 1984 collection of nonfiction. By their titles alone, two chapters in that book—"Uses of the Erotic," "The Uses of Anger"—invoke a rhetoric of tool use in order to describe just what might be instrumentalized toward the dismantling of power. Along with eroticism and anger, desire and emotion, Lorde finds a similar use for literary language itself. Literature has only ever been institutionalized by universities and publishing, and it has only ever contributed to the further institutionalization and perpetuation of these industries. It must therefore be regarded with suspicion, for its near-certain associations with the house and tools of a master. But according to Lorde's 1977 essay "Poetry Is Not a Luxury," also collected in *Sister Outsider*, it may be that literary language is not simply one more of the master's tools. Poetry, for Lorde, may also be the vehicle for any subsequent imagination of feminist solidarity and racial justice: "We can train ourselves to respect our feelings and to transpose them into a language so they can be shared. . . . Poetry is not only dream and vision; it is the skeleton architecture of our lives. It lays the foundations for a future of change, a bridge across our fears of what has never been before."[12] Notably, Lorde's metaphors are again technological. As architecture and foundations, poetry is a way of building. As a bridge, it is a way of connecting, although one delinked from any fantasy of perfected telecommunication. Poetry is a way to talk and build together that is wholly separate from the latest electronics.

Rather than the communicative architecture of a network, Lorde theorizes poetry as a "skeleton architecture" that is both somatic and affective, combining the virtues of eroticism and anger. Rather than flows of power linked across a planet, she theorizes a bridge that leads away from power and toward shared language. As Roderick A. Ferguson writes, Lorde appeals to poetic practices as "the resources for establishing a will to connect, especially in those areas where certain connections were often prohibited."[13] But Lorde's will to bridge is not merely a substitute for telecommunication; it is

a wholly different way of living together, and one that rejects the commandeering of language by industry or the academy. To the charge that poetry
is just another object of a heteropatriarchal inheritance, handed down as the
tool of reactionary institutions, Lorde clarifies: "I speak here of poetry as a
revelatory distillation of experience, not the sterile word play that, too often,
the white fathers distorted the word *poetry* to mean—in order to cover a desperate wish for imagination without insight."[14] The capacity to bridge and
build is not inherent to all poetry, but it does inhere in any poetry that seeks
insight as well as imagination, thus offering access to distilled experience.
These are also the qualities that inhere in Lorde's own poetry. As Ferguson
argues: "For Lorde, poetry was a way to enact an intimate scrutiny needed
for personal *and* social transformation, a way to critically engage the self to
set the stage for new interventions and articulations."[15]

So it is that in 1973, six years before the Second Sex Conference, four
years before "Poetry Is Not a Luxury," Lorde provides her clearest account
of machines and selves, in what approximates a poetic method for political
dismantling. The poem "For Each of You," collected in Lorde's 1973 volume
From a Land Where Other People Live, enjoins its listener to live well and judge
consistently, without distraction by daily realities. "When you are hungry /
learn to eat / whatever sustains you / until morning," the poet writes in the
second stanza, "but do not be misled by details / simply because you live
them."[16] This is a lesson in modeling ethical consumption for subsequent
generations: a detail is part of life, but it may not be necessary for sustenance.
To accept the mere detail as nourishment is to be misled. Lorde expounds in
the next stanza, the poem's longest:

> Do not let your head deny
> your hands
> any memory of what passes through them
> nor your eyes
> nor your heart
> everything can be useful
> except what is wasteful
> (you will need
> to remember this when you are accused of destruction.)[17]

The "details" that the poet mentions earlier are here neither explained
nor enumerated. But they are defined: details are excessive, wasteful. The
addressee—the one whose head must affirm her hands, eyes, and heart—has
no use for them. Everything can be useful, but not everything is. Conversely,
some things are waste, but most things are not. It is for this reason that "you

will need to remember" why you set something aside "when you are accused of destruction." Accused by whom? The poet does not reveal.

She does, however, offer clues as to what separates useful tools from wasteful, misleading details. Memory is useful, as are the listener's eyes and hands. The heart is useful, and the poem itself beats out a rhythm to demonstrate, in the anaphoric repetitions of "nor" and of the short *e* that begins "everything" and "except." Read down the left side of the stanza, nor-nor-eh-eh, the first syllables of these lines beat first like the lub-dub of the heart and then like breath pumped from the lungs. She must let go if she is to preserve what is useful, the eyes and hands and heart. Even so, the end of this sentence is uneasy. The punctuation is wrong, the speaker uncertain. A period falls inside of the parentheses, where grammatically it should arrive afterward. Still, the new sentence begins on the next line of the same stanza, as if the previous sentence had drawn to a satisfactory close:

> Even when they are dangerous
> examine the heart of those machines you hate
> before you discard them
> and never mourn the lack of their power
> lest you be condemned
> to relive them.[18]

This new injunction, which both is and is not a new sentence, is the core of Lorde's ambivalently Luddite response to the institutions and devices that, together, she would later name "the master's house." The word *before* is possessed of a strange ambiguity—it can mark either a conditional or a temporal relation. Here it marks both. On one hand, the word suggests a conditional association of two acts: if you examine the heart of those machines, then you will not need to discard them. On the other hand, it also denotes a temporal difference between the two acts, and therefore intimates a sequence: destroy the heart of those machines, but first examine them. You hate the machines, but you should keep them and examine them. You must let some machines go, but whether you keep them or discard them, you need first to examine them. This is certainly a contradiction, but it is a contradiction that is (to borrow Lorde's own distinction from "Poetry Is Not a Luxury") aligned with revelation and insight rather than sterile play.

This section of the poem returns to the nature of these odious machines, their heart. What are they, these machines with hearts that we hate? They are what is wasteful not useful. They are the distracting details, invoked in Lorde's previous stanza, whose power has now been acknowledged. They are what Lorde's addressee will have to relive, unless she examines them.

They are what she must either retain or discard. The first three lines of the stanza can be read as reframing and reversing a well-known passage in John Ashbery's poem "Illustration," from the 1956 collection *Some Trees*: "Much that is beautiful must be discarded / So that we may resemble a taller / Impression of ourselves."[19] Ashbery's lines mourn the beautiful things that must be relinquished if we are to achieve the greatness that we imagine we possess. Lorde's lines do not mourn anything, and they rest not on the certainty of *so that*, but instead on the discomfiting uncertainty of *before*. The addressee is asked both to preserve and to discard something she already "hates." But first, or in any case, she must examine their heart.

Along with the ambiguity of the word *before*, Lorde's likely nod to poetic tradition has the effect of dedramatizing the act of setting machines aside. By dedramatizing, I mean what Elizabeth A. Povinelli has meant: "We must de-dramatize human life as we squarely take responsibility for what we are doing. This simultaneous de-dramatization and responsibilization may allow for new questions. . . . We will ask what formations we are keeping in existence or extinguishing?"[20] Examine before you discard: with its two antithetical meanings, the line manages not to erase itself, but instead to pronounce an ethical problem.[21] To hold both meanings in mind at once, this is to commit simultaneously to understanding and to relinquishing the hated machine. As in Povinelli's formulation, this means allowing for new questions.

The question that motivates simple technophobia—when, where, and how should I hate machines?—is itself discarded in favor of more particular and contingent questions. In "Poetry Is Not a Luxury," Lorde will construe poetry as the practice best able to produce solidarity, but only if it is somatic and affective and not just a kind of play; only if it is valued for its insight as well as its imagination. In "For Each of You," by contrast, it is the somatic-affective practices—the habits—that must be examined as the site of both usefulness and waste, then dedramatized and responsibilized (to keep Povinelli's neologism) toward ethical survival. There are habits of the body that are useful, affirming the heart and the hands, and then there are habits that are wasteful, interrupting solidarity and denying the value of hands and heart. "Everything can be useful / except what is wasteful": what is wasteful must be examined and then discarded; what is useful must be used. When it is read in light of her later writing about dismantling and poetic possibility, Lorde's ambiguous poem leads neither to the ontological question—what is technology?—nor strictly to the historical question—what have been some particularly advantageous or disadvantageous technologies?—but instead to questions motivated by responsibility: What criteria determine which machines are repaired and retained, and which are "extinguished"? What knowledge can be gained from each machine before its fate? What new questions might proceed from which bits of new knowledge?

In the final stanza of the poem, Lorde shows that developing these questions is a procedure integral to life, and most crucially to black motherhood. They prompt an injunction to:

> Speak proudly to your children
> where ever you may find them
> tell them
> you are the offspring of slaves
> and your mother was
> a princess
> in darkness.[22]

With a closing nod to the title character of *Dark Princess*, the 1928 novel by W.E.B. Du Bois, Lorde shows how dismantling has a specific history and extends toward a specific purpose. The active discarding of machines is not only destructive, but is also a historically conditioned kind of making. Arising from conditions of enslavement, dismantling proceeds through a narrativization of that past in an ongoing practice of historical memorialization. This memorial practice is a pedagogical and parental practice: proud speech to the offspring of slaves as a material and communal form of love.

Dismantling rejects the prelapsarian fantasy of a humanity that preceded machines, and it does not have much to do with the romance of a harmony between humans and machines. It is instead in search of nontechnological, or differently technological, human life in a world full of machines. It is also a call to examine which habits of the body or which clichés in historical narration can contribute to, and which are destined to interrupt, a culture of coalition and mutual obligation. It is an unmaking of reactionary traditions as if they were breakable machines, and a bricolage of the remaining parts into usable histories. Dismantling partakes not only of philosophical approaches to technology, but also of approaches that develop across disciplines, outside the academy, among activists, and in the formal language of literature. Lorde notes that dominant institutions and knowledge formations have been the tools of historical violence. To end that particular violence, therefore, is to break and remake institutions and knowledge formations. To do so, different tools must be applied than were designed and constructed by those institutions. And yet, it remains, you must "examine the heart of those machines you hate / before you discard them."[23]

Epistemological Luddism

According to Langdon Winner, speculation on technology and society should oppose a pair of reactionary strains of thought that had permeated even the more radical philosophical approaches. In his 1977 book *Autonomous Technology: Technics-out-of-Control as a Theme of Political Thought*, Winner called one of these strains the "traditional" account of technology, and the other he called "technological politics." In the traditional account of technology, social formations invent tools for their consequent improvement. On this view: "control is one-directional and certain, leading from the source of social or political agency to the instrument."[24] The second perspective, technological politics, holds no such naive view of technology, as if a machine were something that could simply be employed or refused by those empowered to do so. Even revolutionary theories of technological invention, Winner demonstrates, are motivated by factors (the growth of industrial sectors, the conduct of war, the consolidation of wealth) that exceed the simple fulfillment of social need. Therefore, no matter how radical, technological politics shares a problem with the traditional, instrumentalist approach.

From Marx and Engels to the then-recent work of Jacques Ellul and Herbert Marcuse, Winner notes, thinkers seemed not to be able to think beyond the technological destinies of human life. Traditional theories and theories of technological politics rely on the same conceptual tools and reach similar conclusions. Both endorse a narrative of human history in which, for good or ill, "all persons come to be inextricably tied to systems of transportation, communication, material production, energy, and food supply."[25] As a result, writes Winner, "no matter who is in a position of control, no matter what their class origins or interests, they will be forced to take approximately the same steps with regard to the maintenance and growth of technological means."[26] It is a determinist deadlock.

In the book's last pages, against both tradition and technological politics, Winner proposes a critical procedure through which certain tools may be analyzed, understood, and even set aside:

> Our involvement in advanced technical systems resembles nothing so much as the somnambulist in Caligari's cabinet. Somewhat drastic steps must be taken to raise the important questions at all. The method of carefully and deliberately dismantling technologies, *epistemological Luddism* if you will, is one way of recovering the buried substance upon which our civilization rests. Once unearthed, that substance could again be scrutinized, criticized, and judged.[27]

Winner's critique arises not against instruments of technology but instead against instrumental and technocentric thinking, especially about histories of labor and nation, about the shape of the world, and about thought itself. Winner offers it as "a corrective to the dewy-eyed traditional assumptions about tools, mastery, and endless benefit" that had plagued both the reactionary and the revolutionary views of technology.[28] With the world in a state of reliance on particular kinds of machines, from industry to telecommunication to politics, it remains "to examine the connections of the human parts of modern social technology."[29] What would it mean, Winner asks, to detach a while from technology and from particular technologies, so as to orient social inquiry around the web of human lives and needs, rather than around the unilinear forward march of technological innovation.

In this act of detachment, the thinker/user/citizen (who is also a producer/consumer/subject) becomes like Ned Ludd: the early-nineteenth-century revolutionary, possibly fictional, who inspired a movement of textile workers to smash the automatic looms by which they found themselves replaced. In the lyrics of a Luddite anthem, composed as Ludd's followers were out with their hammers, defending their jobs and their craft through sabotage:

> And when in the work of destruction employed
> [Ludd] himself to method confines
> By fire and by water he gets them destroyed
> For the Elements aid his designs[30]

In Winner's argument that a way of thinking might "recover the buried substance" on which civilization has been built, there are echoes of "General Ludd's Triumph." The epistemological Luddite, like the breaker of looms, must go unconfined to any particular "method" in efforts toward the "work of destruction." Unlike that hammer-wielding hero, however, the epistemological Luddite insists not on smashing, but instead on "carefully and deliberately dismantling."

Pulling Out the Plug

In 1971, in an art magazine called *Avant Garde*, there appear results of a brief survey entitled: "The Machine I Hate the Most." Capitalizing on the anti-technological spirit that had momentarily begun to press back against cosmotechnics, *Avant Garde* solicited brief technocritical blurbs from a cast of public personalities and star intellectuals. In her introduction to the two-page symposium, editor Dorothy Bates proclaimed: "America is witnessing, for the first time in her history, a revolt against the machine. . . . Perhaps it was the incredible expense and frivolity of sending a few men to the moon. Or perhaps it was automation, which has helped throw so many people out of work. Whatever the cause of the revolt, some radicals are even suggesting that we set up a new society of Luddites—named after the 19th-century band of Englishmen who went around smashing machines."[31] Respondents to this prompt included the original Beats Allen Ginsberg and David Amram; downtown hipsters Tuli Kupferberg, a Fug, and Viva, the most famous of Warhol's superstars; two of the Chicago Eight, Mitchell Goodman and Abbie Hoffman; proxies of mainstream Americana like Al Capp and Norman Rockwell; celebrity academics including Claude Levi Strauss and Ashley Montagu; and various others. Even Thomas J. Watson, the board chairman of IBM, managed to submit a reply. In none of the commentary is there even a hint of surprise at the perceived demand for such a survey. Aside from Watson's glibly paradoxical reply ("Thank you for asking me to participate in your anti-machine symposium. It will not be possible for me to do so."),[32] the responses are characterized either by affable good humor or by earnest pleas for technological limits.

Whether or not Bates's "new society of Luddites" had ever been summoned or proposed by any of these celebrities, the answers she gets are implicit admissions that the title question (what machine do you hate the most?) has a certain validity. Some replies are jokes, like the cagey response from the *New Yorker* cartoonist Saul Steinberg—"I can't give you an answer. What am I, a computer?"—or the aesthetic judgment of Isaac Bashevis Singer—who complained of "typewriters that turn out bad stories"—or like the absurdist non sequitur from the pop artist Roy Lichtenstein—"the unicycle." One or two answers are genuinely nostalgic for a pretechnological past, as in Viva's reply: "The ideal world would be one with no more electricity. . . . Before Thomas Edison—that was the best time in history." But most of the responses avoid both the dodge of humor and the lure of nostalgia, and are directed instead at single machines, real and metaphorical, that might truly be lived without. Allen Ginsberg rejects the "conspicuous-consumption machinery" which he compares to a "vast robot monkey on National back." The actress Celeste Holm and the activist Dagmar Wilson agree that the telephone was the worst machine, for it had stunted interpersonal interaction. Other respondents decry sources of noise and air pollution: snowmobiles, outboard motors, supersonic jets, diesel engines, and sprayers of insecticide. Claude Levi-Strauss and Ashley Montagu concur with Philip Berry, the president of the Sierra Club—all three lamenting the disastrous social and environmental effects of car ownership. Levi-Strauss calls driving "a kind of foolish game in which everybody is bound to lose."[33]

Speaking with most urgency were the two members of the Chicago Eight, Mitchell Goodman and Abbie Hoffman, as well as the poet Denise Levertov. All three hate the same machine and use the same phrase to name it: "The War Machine." Refining this phrase, Levertov specifies that it is "bombs, tanks, napalm, nerve gas, etc., etc., etc." that she most hates. Hoffman insists that "we're working on pulling out the plug." Goodman concludes that indeed "it is time for a revival of Luddism—machine-smashing for the sake of survival."

It is novelist Christopher Isherwood who admits to having most hated "*computers*, mostly because of the sickly idolatry that's practiced toward them."[34] As in the rejection of the war machine, Isherwood's critique of computation takes the form of a double gesture. First, technology is a thing in the world. Indeed it can be anything in the capacious category that is easily stretched to include not only beneficial machines, but also machines that destroy the planet, machines that interrupt social relations, and machines that inhibit survival. Second, technology is the object of a critique. It is as an idea, and not only as a class of thing, that technology is fairly criticized. On humanist grounds, technology is a distraction from the path to human freedom. It is the real device that may or may not be worth destroying or discarding, and it is also the imaginary object of a "sickly idolatry" that itself should be destroyed or discarded. Luddism, whether in the philosophical sense of Winner's epistemology or in the more literal sense of the respondents to the *Avant-Garde* survey, is not a disgust with the results of scientific or computational research. It is a commitment to the finding of freedom. The terms of such freedom, and the technique of its expression, were very often literary.

To Lay Bare Hidden Mechanisms

In Quebec in 1976, while on a lecture tour of North America, Édouard Glissant pronounced literature to have two principal functions. One of these functions, he said, was to act as a site of national or racial solidarity, while the other was "a function of desacralization, a function of heresy and intellectual analysis, by which to dismantle the machinery of a given system, to lay bare hidden mechanisms, to demystify."[35] For Glissant as for Lorde, poetic language is a kind of dismantling. Existing power formations are a system with a machinery that it is poetry's function to take apart. Dismantling, then, is a form of analysis that can strip from inherited power formations their mystical and ideological carapace—their mantle. Glissant's specific target is the predominantly occidental fiction of an ostensibly unitary world that, a decade earlier, he had called an "ideational globality."[36] To any such fiction of information that would see transparency as a communicative principle, Glissant opposes his poetics of opacity. Opacity, in Glissant's words, "encroaches on the mechanics, the technologies" of the world, "to *thicken*" them."[37] This thickening, in turn, is at the core of the literary politics of dismantling: a way to take apart and then relinquish thinking that is technocentric and occidental. As one kind or category of dismantling, opacity denies the primacy of free and transparent communication, while jamming the metaphors of an electronic or mechanical world. Dismantling is thus recognizable as an imperative of critical reading, largely because of its employment during the Long Seventies. Much literary theory of that decade aims to prove destructive toward reactionary aesthetic and cultural standards, while also proving productive of tools for continued analysis of these standards. In this it differs from the traditions in literary theory that precede it. But it does not only differ. It also aims to dismantle and then to relinquish those traditions.

In this, Glissant resembles other critics like William V. Spanos and Paul A. Bové, who (sharing Glissant's commitment to revising Heidegger) chart a poetic project whose objective is to take a hammer to reactionary prior practices. Both Spanos and Bové chafed against modernist New Criticism, which had in their eyes produced an exclusively spatial notion of the literary object. When a critic attends only to space and not also to time, they argued, a literary object loses its figurative trajectory or its narrative arc, grinds to a standstill, and ultimately loses its capacity to reflect its relation to the real world. After modernism, both agreed, poetry had begun to foreground its temporal aspect. If dismantling has two discrete senses—to break or take apart; and to remove the mantle, cloak, or outer wall—Spanos and Bové express each sense of the term by blending the poet's capacity to smash conservative literary traditions with the critic's capacity to make criticism that is both naked and unsteady.

Thus Spanos can argue in 1970: "The need to retrieve the exploratory temporal stance, i.e., to destroy the metaphysical circle of the Western literary tradition, is not restricted to contemporary phenomenological philosophy but is also felt deeply by the contemporary, the post-modern poet."[38] And a half decade later (in a dissertation supervised by Spanos), Bové can argue: "Once the critic is stripped of the presuppositions upon which he bases his enterprise, the text and the reader are deprived of stability, and criticism emerges as radical flux, in which the text and the interpretation are constantly modifying, adjusting, and perhaps even destroying each other."[39] Like Glissant, with his metaphors of thickening and taking apart, Spanos and Bové identify tasks for criticism and poetry: to destroy and be destroyed; to strip bare; to counteract or counterindicate the literary tradition; to examine the heart before discarding it.

It is a double line of thinking, utopian and practical, hopeful and skeptical, that gives way to this variety of literary and epistemological Luddisms. This is not to say that the Long Seventies produced sufficient tools for handling technological politics. Indeed, as Langdon Winner himself demonstrated at the end of the era, it was the insufficiency of these tools that led to a need for epistemological Luddism:

> Goodman's plea for the application of moral categories to technological action, Bookchin's outlines for a liberatory technology, Marcuse's rediscovery of utopian thinking, and Ellul's call to the defiant, self-assertive, free individual—all of these offer something. But when compared to the magnitude of what is to be overcome, these solutions seem trivial.[40]

In short, when faced with the hopeful new ways of using technology, or dire new ways of destroying ourselves with technology, it becomes possible to choose a third less-trivial option.

Such a mobilization of the word *technology* invokes central philosophical approaches to the concept, most notably by Marx and Heidegger, but it also sets them aside rather than merely building on them. Perhaps like the books of Glissant, and then Spanos and Bové, *Autonomous Technology* could only have been written in the Long Seventies. The problem was, Winner admitted, that "there have never been any epistemological Luddites."[41] Yet had there not been? Is not epistemological Luddism in fact a species of relinquishing or dismantling that can be seen across the period? Whatever it is called, this is a critical attitude toward the machines and thought formations that spill out from the master's house. Extending from activist to literary to philosophical discourses, epistemological Luddism asks what machines even are, and what it is we want from them, even if they are destroyed in the process. In its conjunction with literary practices, epistemological Luddism has meant two things: one, it has meant broadening the scope of literary analysis to encompass technology as well; and two, it has meant wielding literary knowledge in opposition to any practice that is technocratic or machinic.

The most literal Luddism of the Long Seventies is likely that which took root in the protests against the war in Vietnam. Taking aim at corporate and cultural proxies of the U.S. military, peace activists frequently used sabotage as a way to slow the war machine, to bring media attention to the struggles of colonized people, and to express solidarity with a planetary antiwar movement. Even within literary culture, this Luddism found expression, most notably in the actions of H. Bruce Franklin in opposition to the military research of his employer, Stanford University. Franklin, an associate professor of English at the end of the 1960s, was a scholar of Herman Melville, science fiction, and the black radical tradition. He was also a Marxist, with intellectual and political investments in the written work of Mao, Stalin, and other revolutionary thinkers. As a leader of the Northern California leftist group Venceremos, and as a tenured member of the university faculty, Franklin led discussions and protests against Stanford's explicit and material support of U.S. empire. These protests culminated at Stanford's computer center on February 10, 1971, in a student occupation that led to work stoppage, unplugged computers, and consequently lost data. This occupation, along with related activities, would lead Stanford to fire Franklin, in what became one of the best-known cases of tenure revocation in the U.S. academy.

Explanations for Franklin's firing are varied, but all rely on claims that Franklin's teaching was (in the words of the review board) "uncomfortably heterodox,"[42] and that his part in the occupation of the computer center had exceeded the remit of his academic appointment. In the rhetoric of Franklin's dismissal, there is a tone common to indictments of purported terrorism, including an insistence that the occupation was a violent act. However, as one speaker explained, during a rally the previous evening, the refusal of violence was the whole point of the occupation: "The beauty of the computer center was no violence is necessary. None at all to get, to get what you want. Simple little sitting in doorways with a thousand people, you know, no violence."[43] This is not to say that Franklin's tactics were peaceful. Indeed, Venceremos resembled the era's other revolutionary groups in its insistence that armed struggle may one day become necessary if the expansion of empire and the application of state racism are ever to halt. Still, the computer center occupation was not violent and was a strategic act of machine-breaking in which nothing—nothing aside from data—would need to break. As the day of the protest dawned, Franklin shared his comrades' diminishment of the need for violence. Because "blood brothers and sisters of us [are] killed in Laos, in Vietnam and Cambodia, in the black and brown communities in the United States of America," he explained in speeches on February 10, it then becomes necessary to "inconvenience ourselves a little bit and begin to shut down the most obvious machinery of war, such as, and I think it is a good target, that Computation Center."[44] To fill the computer center with the living bodies of scholars, from undergraduate to faculty, was to perform an inverted metaphor for the accumulated deaths in Southeast Asia. It was to produce a small amount of inconvenience for a great deal of noise. It was to name Stanford's research apparatus, its computers, as "the most obvious machinery of war." And it was, even without smashing, to silence that machinery.

Gradual Painless Relinquishing

"The Remembering Machines of Tomorrow" first appeared in *The New Yorker* and was later published in Merwin's 1969 volume *The Miner's Pale Children*. In that book, Merwin's words about a relinquished device sat alongside other prose pieces that are equally epigrammatic, or else largely mystifying. Robert Scholes writes of the volume: "These bits of prose are magical; they only *look* more digestible than poetry. Once taken, they grow inside the reader: first filling him comfortably, contenting belly, heart, and brain, then expanding beyond comfort, forcing the eyes toward new perceptions, straining the ears toward unheard questions."[45] Merwin's prose pieces, when they take up the invention and relinquishing of technology, are not Bible stories with fabular lessons. Rather, they "force the eyes" and "strain the ears" of a reader who has grown comfortable and contented with contemporary media. When Merwin poetizes the "remembering machine," it is not a prediction of technologies to come, as if he had somehow divined the possibility or the obsolescence of smartphones. Instead, in the moment of its writing, the remembering machine is properly literary: a poetic "instrument" and not something that can be built. Neither a database nor a databank, as might be described in a work of so-called hard science fiction, the remembering machine is an imaginative device and narrative artifice: a widget that remembers, and that precipitates, in Merwin's language games, a reimagination of human behavior at the dawn, and then the dusk, of a technical revolution.

It is in this last stage, the dusk, the fading of the second machine age, that Merwin's vision most resembles the visions of other epistemological Luddites. Like Lorde, Winner, Glissant, and the others, Merwin is more interested in the social and cultural expression that surrounds media change than in the technical makeup of the particular media. In Merwin's science-fictional mode, thinking machines are not the achievement of human futures. They are instead what provides opportunity for imaginative commentary on the present. They are not real, but only exist as a figure. If Scholes is correct that Merwin has disturbed the human subject, forcing the eyes and ears, then the disturbance occurs through a transfiguration of human faculties for memory and hope.

In Merwin, the new technological order has led to new conditions of memory. When memory can be farmed out to a machine, it loses its status as a guarantor of history or identity. Memory fades, in "Remembering Machines of Tomorrow," because it has been tasked to a computational device, the machine of its title, which is itself then abandoned. In an earlier chapter of the book, entitled "Memory," the limit of recall and remembrance is not the technology of the machine, but rather the persistent fact of a spatiotemporal present: "Here was something he had not remembered. . . . What could he call it? His presence in the place? The standing on the needle? The present? The blankness at the end of the story?"[46] The paradox of memory is that it exists in the present on behalf of the past. It stands "on the needle"—of what? a compass? a phonograph? a thread?—and so retains nothing of its own moment, its own present. It is a blankness. Its poverty is its failure to include what is most immediate to it. The technical prosthesis, that is the "remembering machine," cannot overcome this evident problem. The machine can accommodate "the whole of his past—or what he takes to be the whole of his past," but that is all; only "his attachment to it will constitute the whole of his present—or of what he takes to be the present."[47] Absent from memory, the present is a blank site of contestation and a reservoir of the user's desire, enthusiasm, and "attachment."

As for forgetting, Merwin takes it up in a chapter aptly called "Forgetting," a few dozen pages after "Memory." There, the limit of memory is neither technology nor the blank present but instead the amnesia of its title: forgetting, but not the sort that would leave the rememberer intact. Instead, Merwin makes amnesia a matter of the world's final dissolution. "Each of the senses at the same time," writes Merwin, "reveals to us a different aspect of the kingdom of change. But none of them reveals the unnameable stillness that unites them." Memory is here an accumulation of sensory connections, and yet it remains broken by a disunifying stillness. There, "at the heart of change it lies unseeing, unfeeling, unhearing, unchanging," but: "None of the senses can come to it. Except backward."[48] The world of remembering machines is a changed world, while also changed are the senses by which the world can be apprehended. As a result, the fullness of memory, defined as the linkages among sensory data, can only be felt after the change has ceased, "backward," from the end of the world. This is indeed where "Forgetting" concludes, two short pages later, when: "Earth has gone. We float in a small boat. . . . Then the water is gone and there is only the small boat floating in nothing in the dark. . . . Then nothing has gone."[49] The disunifying stillness, once occupied, can pull senses together into a dynamic memory, involuntary memory, but only after it is too late, when even the "nothing has gone."

Inasmuch as Merwin establishes memory as the field of agonism in a present of technological change, his book also establishes these very terms—present, technology, and change—as ideas to be negotiated in pursuit of any nourishing form of memory. The disunifying stillness is the space of dedramatization, in Povinelli's sense, where universalist presumptions and apocalyptic exaggeration are set aside for better, more specific questions. This is the space-time where hard work might proceed, whether that is the hard work of remembering, for Merwin; of solidarity in coalition, for Lorde; of recovering and scrutinizing "the buried substance," for Winner; of dismantling a given system, for Glissant; or of a text, for Spanos and Bové. Luddism, when it is this sort of relinquishing, is not about machine-smashing at all. Or rather, it is not about smashing every machine, and it is not about smashing any machine before it is examined.

CHAPTER 2

Communion

all the hollow forms of communication and "togetherness" . . .
lack real communion or real sharing

—Ursula K. Le Guin, "The Child and the Shadow"
(1974)

To destroy machines, or even just to examine
and relinquish some of them, is a risk. In a cosmopolitan world whose cos-
mopolitanism often seems a side effect of its telecommunicative and travel
technologies, it is fair to suggest that we (even if not a reflexive and hegemonic "we," but instead the coalitional "we" of a planet at cross-purposes;
the "we" of a variegated species; or the "we" of a multispecies assemblage)
might need a few those machines to live. Or, more precisely, that we need
them to live together. This way of thinking is owed partly to the metaphors
of cosmotechnics that were linked to one another in the mid-twentieth century, and that extend, rarely questioned, into the present moment. This way
of thinking is also profoundly destructive, contributing to an overreliance on
machines, feeding a fatalistic acceptance of new technological constraints
on life, and lending itself to habits of ideological speech. In such speech are
superimposed dissimilar ideas: language is accepted as a kind of medium,
any medium as a message, the world as a network of communication, acts
of communication as ways to cohabitate in the world, global practices of
cohabitation as variations on familiar metaphors like Marshall McLuhan's
"global village" and Walt Disney's "small world after all," and R. Buckminster Fuller's "Spaceship Earth." The ascendance of these superimpositions is
partly owed to the still-rising dominance of the global satellite and computer
systems that they invoke. Yet the metaphors also remain metaphorical, and

are therefore disputable, both in their existence and in their ethics. They are metaphors of machines, the connective systems of cosmotechnics, and they are metaphors as machines: clichés that resemble technology to the precise degree that a writer may tinker with them, dismantle them, alter them, direct them toward a different purpose, or replace their mere description of global proximity with an urgent ethics of planetary communion.

To Live on a Small Globe

Is it even possible to live and work in the world together? Or put differently, what words for the world might stem from an imagination of collective life and work? *Cosmotechnics* was coined by Albert Szent-Györgyi, an émigré public intellectual and an opponent of nuclear weaponry. For Szent-Györgyi, in his 1971 pamphlet *What Next?!*, the dawn of networks invokes not only a new historical phase but also a new ethical attitude toward the world. As a period in the history of machines, the cosmotechnic present differs from previous periods in the history of machines, the periods of neotechnics and paleotechnics, as these were marked out by Lewis Mumford's *Technics and Civilization* in 1934. "The paleotechnic phase is a coal-and-iron complex," Mumford had written, "and the neotechnic phase is an electricity-and-alloy complex."[1] To these phases, Szent-Györgyi appends cosmotechnics, a period in which the world has come to be characterized by a linked complex of satellites, cables, and bombs.[2] A cosmotechnic world is a world in which words and weapons might arrive at their destinations with ever-greater speed and effectivity. This effectivity is what leads Szent-Györgyi to move past mere periodization toward an ethics of technology, writing: "War, in the age of cosmotechnics, has its specific features. Earlier, if two nations quarreled they could fight it out between themselves. . . . War in the cosmic age is different: the bystanders will be killed too, and all mankind will be wiped out."[3] Even under the advances of electricity and alloy over coal and iron, Szent-Györgyi argued, military conflicts could damage only those who had decided to engage in war. Now with the dawn of military and civilian networks, the whole planet was newly endangered.

In Cold War binarism, Szent-Györgyi betrays an ignorance of colonial warfare and indeed of any bilateral warfare that is entered involuntarily by at least one side. Even still, he marks a sensible distinction between the technologies of small-scale and large-scale death. The consequence of cosmotechnics, Szent-Györgyi concludes, is that something has to change: "It is not possible to live on a small globe with hatred and fear in our hearts and megaton bombs in our hands."[4] Assuredly, Szent-Györgyi accepts that there has been a real change in the scale and traversability of the world. Yet unlike the futurists that were his contemporaries, Szent-Györgyi's interest is in the propagation of peace over and above the propagation of machines. Throughout the work of McLuhan and Fuller, say, there is a perceptible faith in what the nuclear physicist Alvin M. Weinberg, in 1967, would advocate as a "Technological Fix"—in short, any scheme by which "social problems [might] be circumvented by reducing them to technological problems."[5] For Szent-Györgyi, by contrast, technical solutions are only ever stopgaps against subsequent real solutions that would inevitably still be required. In this way, to war, "the technical solution lies in a strong UN [but] the real and final solution is a spirit of human solidarity"; while more broadly speaking, "our technical solution to *violence* is tear gas and the National Guard" but the only real (i.e., not only technical) solution will be to loosen "the rigidity of attitudes which will budge only to active violence, begetting it."[6]

What differentiates prior thinkers from Szent-Györgyi is what differentiates capitalists from communists, a fact that Szent-Györgyi, a communist, admits and explores. But more fundamentally, the difference is between thinkers who accept the technological fix and a thinker who does not, insisting instead that "technical solutions, in themselves, may not only be ineffective but may achieve the opposite of what they are designed to achieve—force creating counterforce."[7] What, in the history of everything known, would make anybody think that a future planetary community, shimmering in its supposed transcendence of violence, might ever be accomplished by technical means? What would make anybody think it could be accomplished at all? The answers, like Szent-Györgyi's appeal to the spirit of human solidarity, are all clichés. Yet if bad clichés can lead to better clichés, maybe some collective form of peace may eventually be imagined. Cosmotechnics is as good a label as any, at least in the historical and ethical sense framed by Szent-Györgyi, for clichés of a planet wired by new technologies and shrunken down by diminished resources. Yet beyond cosmotechnics, and therefore beyond Szent-Györgyi, what better clichés may exist?

In a world where it is assumed that there is not enough to go around, communication technologies are said to help humans share what little is available. Yet, according to another economics, the economics of abundance, planetary resources are not at all inadequate, but have simply been underproduced, hoarded, and jealously guarded by a greedy few. This latter economics involves a critique rather than a valorization of communication technology. It calls on anarchism and Marxism rather than capitalism. From the economics of abundance comes a notion not of commons, as the pool of shared resources in whatever quantity, but rather of communion, as a collaborative practice of making do. Communion is a kind of dismantling, inasmuch as it requires a break with, rather than a mute acceptance of, the clichés of a cosmotechnic world.

Riders on Spaceship Earth

In a way, networks and cosmotechnics have their roots in the very claim that the world is round, traversable, and whole. Yet contemporary thinking about networks really begins in the middle of the 1960s, in the years that followed the launch of the Telstar 1 satellite in 1962, and then the Intelsat I satellite in 1965. It then radically accelerates with the establishment of computer networks, particularly after the launch of Arpanet in 1969. In the popular imagination, satellite and computer networks unified a formerly diffuse world under the sign of global connectability. The fact that these new technologies were geared toward military advantage and private profit (Telstar was a property of AT&T; Intelsat was owned by Comsat; and Arpanet was a tool of the U.S. Department of Defense) did little to dampen their utopian association with the public good. These were networks, and networks seemed to bring the world together, and living together seemed mostly to be better than living far apart. Television programs, like radios and telephones before them, at last appeared to connect the patterns of consumption in one home to the patterns of consumption in all the other homes in the same broadcast radius. When we watch the same shows, it follows, and can call each other on the phone at any time, we start to have something in common. This something-in-common, in turn, is the discrete set of values that are said to compose a media culture. A popular and scholarly language develops, through which to debate the positive or negative influence of those values; and yet the fact of those values, along with the networked mode of their distribution, is rarely questioned. The more that cables and satellites transmit their communicative signals, the more completely can a network of stories and values seem to stretch, like a net or a web, over the whole planet.

The network, however, is not just how things are. It is a metaphor, initially inscribed in the language of technological boosterism. In 1970, *Evergreen Review* published an article by R. Buckminster Fuller entitled "Man's Total Communication System," with the lede: "Humans are emerging from the womb of Space Vehicle Earth which took care of our needs in the past. Now we must operate our planet with the full understanding that our life-support system has no more room for error."[8] From nothing but its title and its lede, the article already associated the metaphor of networks with the full maturity of the human species, and joined both to the economics of scarcity.

New communication technologies had been sent high in the air, deep under-ground and underwater, even as there remained starving people in the world. For a writer like Fuller, it seemed only natural that one thing, networks, had something to do with the other, scarcity. For him, this moment was the culmination of human evolution. Anything occurring beforehand, that is before networks, must have happened in the "womb" of the ship, so that the birth of human society is simultaneous and coterminous with the dawn of the network: man's total communication system. Human beings crawl from the ship's womb into its cockpit, its storage and cultivation facilities, and its living quarters. But Fuller cautions: "Our life-support system has no more room for error." In short, too many mistakes have already been made in the depletion of food and water supplies and in the damage to the planet; this old/new world must be carefully maintained if human beings are to survive.

Were Fuller to have stopped there, with only a title and a lede, he would already have defined his position only in terms of an unsustainable econom-ics of scarcity. He draws a connection between communication and resource management that is unintuitive and too programmatic. Moreover, in the remainder of the long article (with illustrations, speculations, improvisa-tions, and a closing column of verse) Fuller expands on this economics as if it were already given. He meditates on the scene of birth, as humanity emerges onto a planet fully wired by communications technologies. "As each successive child is born," Fuller writes, "it comes into a cosmic conscious-ness in which it is confronted with less misinformation than yesterday."[9] This decrease in misinformation is what makes life possible at every level in a syn-ecdoche of scale: individual, species, universe. For Fuller: "Man is a self-con-tained, micro-communicating system. Humanity is a macro-communicating system. Universe is a serial communicating system. . . . Communications are at all times invisibly present everywhere around our planet. They permeate every room in every building—passing right through walls and human tis-sue."[10] The image is of a species that could not formerly have been what it is now, at the individual, collective, or universal register.

Before new devices began to send and receive their signals through the flesh of bodies and structures, there could be nothing properly human, but only a species in its latency. Of space travel, taking place as it did before the new communication system was established, Fuller can only conclude that it was "tantamount to a premature, temporary surgical removal of a baby from its human mother's womb."[11] Spaceship Earth is thus not only the simultane-ous birth of the species and the network, but also the birth of the universe and the establishment of a regimen by which to manage finite but renewable resources. Here as elsewhere, Fuller is drawing on observations in his book

Operating Manual for Spaceship Earth, published two years earlier. There, laying groundwork for the later article, he writes: "Before the invention and use of cables and wireless 99.9 percent of humanity thought only in the terms of their own local terrain. Despite our recently developed communications intimacy and popular awareness of total earth we, too, in 1969 are as yet politically organized entirely in the terms of exclusive and utterly obsolete sovereign separateness."[12] Fuller's notions of communications intimacy and total earth are, as much as McLuhan's global village, privileged models of the cosmotechnic imagination. There has been actual change in the scale of the world, as far as Fuller is concerned, and until this change is fully accepted by the rest of the species, humanity is bound to get wrapped up in outmoded forms of separateness. There is something postnational about Fuller's perspective, but it is not the left postnationalism of communism or anarchism. As gestures toward borderlessness go, it falls much nearer to the imagined "free flows" that define capitalist globalization. Not the insistence on a network of workers, or of communities, it is instead a faith in the autonomous conduct of the total communications system. National boundaries fade, on this view, because they get in the way of rapid message transmission, not because they are unjust.

The idea that the earth is like a spaceship arose partly as a way to make sense of the emergent global telecommunications and computational network; and partly it just amplified that idea, as the technological solution to the problem of resource management. It was a myth, and a persistent one, that fueled research and development just as it influenced global political policy. When Fuller's operating manual went mainstream, it became a shibboleth of tech culture and political optimism. But before then, Spaceship Earth was already a conceptual tool of liberal internationalist policy.

Preserved from Annihilation

On July 9, 1965, Adlai Stevenson addressed the Economic and Social Council of the United Nations. Stevenson was at that time the U.S. ambassador to the UN, a body he had helped found, and he was to die five days later of heart failure. Charting a way forward for the UN, in the face of an advancing technomodernity and at the dawn of the U.S. war against North Vietnam, Stevenson said: "We travel *together*, passengers on a little space ship, *dependent* on its *vulnerable* reserve of *air* and soil; *all* committed for our *safety* to its *security* and *peace*; preserved from annihilation only by the *care*, the *work*, and I will say, the *love* we give our *fragil* [*sic*] craft."[13] This image of the human collective, beset by challenges, limited in resources, is even more precise than

Fuller's later formulation, in its equation of "a little space ship" with the hard problem of world community in an economy of scarcity. For Stevenson, the metaphor of Spaceship Earth was not merely a futurist vision. It was an internationalist project.

Moreover, as Stevenson told the assembly in his final speech, it was an internationalist project best suited to the United Nations: "*Joint* action is . . . *still* held back by our old parochial *nationalism*. We are still *beset* with dark *prejudices*. We are still *divided* by *angry*, conflicting *ideologies*. Yet all around us our *science*, our *instruments*, our *technologies*, our *interests* and indeed our deepest *aspirations* draw us more and more *closely* into a *single neighborhood*."[14] For Stevenson, Spaceship Earth was a dream, a plan, and an aspiration. Nationalism, a form of "dark prejudice," was the obstacle that an international organization, bringing aspiration to bear on science and technology, might overcome. Because the spacecraft is so "fragil," Stevenson concluded, "we cannot maintain it *half fortunate, half miserable, half confident, half despairing, half slave*—to the ancient enemies of man—*half free* in the liberation of resources."[15] Spaceship Earth was a way to picture life after the resolution of these internal tensions, which in turn Stevenson judged to be the task of the UN: the end to exploitation and the dawn of real egalitarianism. Coined in the late nineteenth century by Henry George, and most closely associated with Fuller, "Spaceship Earth" nevertheless got its contemporary meaning from this speech by Stevenson. For Stevenson, a former Illinois governor and two-time presidential candidate, Spaceship Earth is a perfectly sensible, practicable, and liberal vision of inclusion and mutual care. Yet there are two main problems with his vision.

For Stevenson, as with Fuller, the first problem with Spaceship Earth is the logic of scarcity that is at its base, a logic that was too easily exploited toward ends that were far more selfish than they were communal. For Stevenson, peace was only to be achieved through the "liberation of resources" from those who would control them. For Fuller, less invested in peace, and more excited about the nascence of a mature version of humanity, resource management was really aimed at the preservation of this delicate new species. Both faced an insuperable contradiction insofar as the U.S.-based network of communications and commerce had only ever been ungenerous in the distribution of global resources—having buried the tendrils of its media and culture deep into the global economy, then having applied that economy and that culture toward the task of further extraction. This contradiction not only undercut the seemingly magnanimous metaphor of Spaceship Earth, but also gave way to metaphors of scarcity that were racist and nativist.

The most dangerous of these nativist metaphors, Garrett Hardin's "lifeboat ethics," can only be thought in relation to Spaceship Earth. Hardin rose to modest fame with his 1968 essay "The Tragedy of the Commons," which has become a standard text in the economics of resource scarcity. He later returned to this topic in 1974 in an essay entitled "Living on a Lifeboat," which responded to the increasingly dominant ideal of Spaceship Earth. For Hardin, "the spaceship metaphor is used only to justify spaceship demands on common resources without acknowledging corresponding spaceship responsibilities," chief of which is the need to constrain immigration.[16] The world is not a capacious spaceship, says Hardin, but just a collection of little lifeboats that cannot accommodate any new passengers. Some lifeboats are comfortable and rich in resources while others are overcrowded and poor, Hardin argues, which makes the guiding question of global technological ethics: "What should the passengers on a rich lifeboat do?" Hardin's answer is that the rich must be selfish when "the poor fall out of their lifeboats and swim for a while in the water outside, hoping to be admitted to a rich lifeboat."[17] Hardin's cruel image for the commons is opposed to the inclusive image of Spaceship Earth, at least on its surface. But both metaphors assume that the world is impoverished by an inexorable scarcity. Against this merely partial opposition, there remain entirely other ways of thinking about the commons (from Elinor Ostrom to Michael Hardt and Antonio Negri to Fred Moten and Stefano Harney) that may yet avoid the slide from scarcity to rapacity. After all, in neither the spaceship nor the lifeboat will there ever be enough food, water, and space for everybody to board. From both the spaceship and the lifeboat, someone gets left behind.

The second problem with Spaceship Earth, especially in Stevenson's vision of it, was also with the United States itself, which had lately begun an invasion of the Dominican Republic and a campaign of reprisal bombings in North Vietnam. It is unlikely that Stevenson knew much about the latter incursion, since the full extent of U.S. covert activities was held secret until the leak of the Pentagon Papers during the Nixon administration. But what was clear to the activist public would also have been clear to Stevenson: the U.S. military encroachment, like its cultural encroachment almost everywhere else, was designed primarily to contain the ideological threat of Soviet communism. With this increasingly common knowledge came discontentment among political and cultural radicals over Stevenson's continued willingness to serve Lyndon Johnson's administration. Between the beginning of the Dominican intervention in late April and the Spaceship Earth speech (and its author's subsequent death) in early July, there had been multiple calls for Stevenson to resign. One of these was signed by an array of leftist critics and writers:

> We have watched in dismay as our government—by its actions in Vietnam and the Dominican Republic—has clearly violated the United Nations Charter, international law, and those fundamental principles of human decency which alone can prevent a terrifying world-wide escalation of suffering and death. . . . We urge you to resign as United States Ambassador to the United Nations, and having done that, to become spokesman again for that which is humane in the traditions and in the people of America. By this act, you can contribute immeasurably to the prospects of world peace. By remaining in your post—without speaking truth to power—you will have diminished yourself and all men everywhere.[18]

The signatories—including Dwight MacDonald, A.J. Muste, Paul Goodman, the novelists Kay Boyle and Harvey Swados, and the cultural critic Nat Hentoff—had caught Stevenson in an ethical failure to rise to the standard that he had himself set, during his presidential campaigns and in his coauthorship of the UN charter. The Spaceship Earth speech, given a couple of weeks later, is therefore not only the central metaphor of liberal internationalism. It is also Stevenson's last-ditch effort to bind that internationalism to an economics of scarcity. This is how Stevenson saw the United Nations as capable of something the United States could not achieve: peace, environmental protections, and social responsibility through the transcendence of nationalism and class.

At the same time, Spaceship Earth was never separate from, but was always a component in, U.S. administrative efforts at moving beyond mere containment of communist threat. Writing back to Goodman in the wake of the call for his resignation, Stevenson defended his "hope of transcending the static policy of 'containment' and moving to the more creative tasks of building a world security based on law";[19] by contrast to leftists like Goodman who "may believe that Communist powers are not expansive" or that "the changes they seek to support by violence are beneficent changes which can be achieved by no other route."[20] Stevenson's letter to Goodman was never sent, and in any case is thought by some to have taken "too hard a line" to have been written by Stevenson, especially in the last days of his life. It was therefore probably written by Stevenson's close friend Barbara Ward, a British economist. Ward's own book *Spaceship Earth* was to be published in the following year, and it struck similar tones, by repeating the familiar themes of technologically enabled togetherness: "Modern science and technology have created so close a network of communication . . . that planet earth, on its journey through infinity, has acquired the intimacy, the fellowship, and the vulnerability of a spaceship."[21] But Ward likewise echoed familiar images of the exclusion of political difference, saying of communism that "to suggest a truce with this evil force is to betray the millennium. There can be no peace, no brotherhood, until these serpents have been expelled from mankind's potential Eden."[22] Reading Stevenson with Ward as well as Fuller, and with critics on the Left and Right, it becomes clear that the metaphor of Spaceship Earth should only ever be read as an enthusiasm for emerging technology, a tactic of anti-communism, and a rationale for excluding the already poor from the distribution of vital resources. It was a metaphor, after all, whose model of inclusion served as an implicit exclusion of anybody not already committed to the capitalist management of scarcity.

It is because the world is "a network of communication," wrote Barbara Ward, that it can be metaphorized as an intimate spaceship. Even aside from Fuller and Stevenson, aside from the metaphors of spaceships and lifeboats, the metaphor of networks weighs like a burden on critical discourse. Media theorist Alexander R. Galloway has noted the infectious quality of this metaphor, such that when one "still believes the old myth that 'networks are enough,'" one tends also believe that "systems are enough to disrupt hierarchies, that networks corrode the power of the sovereign, that markets are the most natural, most democratic, and most scientifically accurate heuristic for redistributing and indeed defining knowledge."[23] To embrace networks is to believe that revolution is already under way and that resources will eventually find their way to those who require them. Opposed to such a perspective, writes Galloway, we should "acknowledge the historicity of networks . . . [and] acknowledge the special relationship between networks and the industrial infrastructure, a relationship that began in the middle of the 20th century and has become dominant."[24] To historicize networks, and so to dismantle the spaceship-like ideas of togetherness that rely on networks, is to ask questions that diminish their control by demonstrating that they are a fiction. What are some of the ways that the metaphors of world-as-network and network-as-world-making have been amplified and contested? And what alternative metaphors might posit other ways to think about the shared planet?

Real Coexistence

The communicative economy of scarcity, as celebrated in the cosmotechnic metaphor of Spaceship Earth, is opposed by a science-fictional critique of that economy by Ursula K. Le Guin and Samuel R. Delany, among others.

At the Library of Congress in 1974, Le Guin lamented that "real community" fails when people subordinate their self-understanding to "the mass mind, which consists of such things as cults, creeds, fads, fashions, status-seeking, conventions, received beliefs, advertising, popcult, all the isms, all the ideologies, all the hollow forms of communication and 'togetherness' that lack real communication or real sharing."[25] The *mass mind* is Le Guin's name, borrowed from Carl Jung, for the new telecommunicative forms of proximity. It is also a serviceable synonym for Spaceship Earth, or for Ward's "network of communication," which are the "hollow forms" that, for Le Guin, can only lead away from communal practices because they lead toward overvalued machines and away from self-knowledge. Long-distance communication and information retrieval, networks enabling uneasy peace and moderate consumption in a world of scarce resources, are a poor model for living together in the technological present. Criticism of the notion has not always been easy to hear, however. Even where ideals and practices of communication are admitted to be imperfect, they are rarely understood as ideological. Even where satellite and computer networks are seen as tools for surveillance and metadata collection, they are still seen equally as tools in the conduct of diplomacy and the sharing of resources. What notions of the planet might then do without the hope for perfect communication and the reliance on networks, but without giving up on lasting peace? There is something science-fictional about the technologies of message transmission and the metaphor of the planet as a spaceship. It should not therefore be surprising that alternate ideas of human communion are to be found in science fiction of the Long Seventies.

Beginning in 1966, with the novels *Rocannon's World* and *Planet of Exile*, Le Guin founded a fictional universe on the simultaneous emergence, and mutual reliance, of an intercultural institution and a machine for instantaneous message transmission. The narrative of the series, usually referred to as the Hainish cycle, is broken up and cobbled together through short stories and novels that are neither chronologically continuous nor contiguous in space. Among these are some of Le Guin's best-known works, including *The Left Hand of Darkness* (1969), *The Word for World Is Forest* (1972), "The Day before the Revolution" (1974), and *The Dispossessed: An Ambiguous Utopia* (1974). Along with contemporaneous work by writers including Delany, Octavia Butler, Joanna Russ, Marge Piercy, Thomas Disch, Philip José Farmer, and others, these novels and stories embody a new stage in the history of the genre. Whereas science fiction had long experimented with narrative form, and often speculated about the political realities of race, class, and sex, it was not until the Long Seventies that so many science-fiction writers would systematically seek formal and generic solutions to political questions. For Le Guin, the main question was the same one that so concerned Fuller and Stevenson: the question of how to live together.

In Le Guin's cycle, intergalactic telecommunication is accomplished through a device called an ansible, while peace is accomplished at the same scale through the activity of a League of All Worlds (a fictional proxy for the League of Nations) and then by its successor organization, the Ekumen (a proxy for the United Nations). Yet throughout the series of books and stories, the ansible and the Ekumen (or the League) are really just the preconditions for a political analysis that rarely includes them. If the science fiction of the Long Seventies can help dismantle the metaphors of networks and spaceships, then it is partly because it sidelines the very institutional and telecommunicative technologies that it invents. Instead of fictionalizing them in a way that Siegfried Zielinski would call media-explicit, in other words, Le Guin approaches communications and internationalist institutions as media-implicit problems—as focal points for a detailed but largely nontechnical discourse on ethics and politics. Le Guin's cycle draws its name from the people (the Hainish, those who come from the planet Hain) who founded the Ekumen, while it is most remembered for its coinage of the word *ansible*. The ansible has served a similar function in texts across the genre in the decades since, but Le Guin's focus lies elsewhere than in the particulars of organization or invention. Instead, enabled by those procedures but moving well beyond them, Le Guin tugs at the tangled relations between sexual and political economy, and between colonizer and colonized, as between anarchism and capitalism.

The central book of Le Guin's cycle is *The Dispossessed: An Ambiguous Utopia*, inasmuch as it is this book that most vividly concerns the high costs and meager benefits of centralized power and property ownership. The novel tells of twin planets, each regarding the other as its moon: Annares, egalitarian and anarchist but barren and hermetic; and Urras, prosperous for some but devastating for those its economic system has impoverished. The protagonist is a scientist, Shevek, who leaves Annares for Urras so that he may develop his latest theory, the mathematical underpinnings of the ansible itself. In spite of this centrality of Shevek's theoretical invention, the main conceptual problem is neither communication nor technology. Focusing on the difficulty of mutual obligation—of communion—the book instead emphasizes the differences between the two planets. At another time, the book's anarchism, as well as its development and then sidelining of technological developments, would have marginalized it. Indeed not all readers were favorably inclined, like Philip K. Dick, who, in a 1976 radio interview, said of Le Guin's work:

"I really don't understand it. Her whole body of writing seems to me to be like a sermonette . . . gussied up with a kind of literary style. But it's all from the Poli Sci department of the University of California in Berkeley as far as I can make out when you strip the style away."[26] Even so, the book was seen by most critics, even those with strong reservations, as one likely to enlarge and improve the genre. Joanna Russ, for one, was unimpressed by Le Guin's gender politics in her review for *Fantasy and Science Fiction*, but admitted: "I carp because the book has earned the right to be judged by the very highest standards."[27]

The most influential contemporaneous response to *The Dispossessed* came from Fredric Jameson, for whom Le Guin exemplifies "world-reduction," a literary principle by which "what we call reality is deliberately thinned and weeded out through an operation of radical abstraction and simplification."[28] With this procedure, writes Jameson, Le Guin can make Urras stand for capitalism and make Annares stand for anarchism. The problem with Jameson's reading is not that Le Guin does not simplify her worlds. She does. The opposition between the political economy of Urras and that of Annares is the central drama of the book. However, the technique of world-reduction is not itself all that simple. It is in part a ruse of the novel to announce itself "an ambiguous utopia," and then to describe two worlds, neither of which is particularly desirable. It is true that Urras is a stand-in for the forms of consumption and impoverishment associated with capitalism and individualism in the United States. And it is equally true that Annares is obviously preferable, an egalitarian society that rejects property ownership and refuses the violent application of power. For Jameson, the difference between the two worlds is clear: "the device of world-reduction becomes transformed into a sociopolitical hypothesis about the inseparability of utopia and scarcity," such that descriptions of Annares are articulations of utopian thought, while the descriptions of Urras pose "a powerful and timely rebuke to present-day attempts to parlay American abundance and consumers' goods into some ultimate vision of the 'great society.'"[29] Yet while Annares is certainly preferable to Urras, it is also barren, closed to the rest of the universe, jealous of its own stated values, famine-prone, and strict. If Annares is utopian, in short, then it is only in the strict, etymological sense. It is no-place and no place resembles it.

When Jameson writes that Le Guin's utopia is inseparable from scarcity, by contrast to the material abundance of capitalist Urras, he casts Annares as a kind of Spaceship Earth. Annares is preferable to Urras, from that perspective, because it has fewer resources and therefore runs less risk of unequal accumulation. Workers on such a planet are conditioned by this scarcity, duty bound to develop their obligations to the commons and to each other. However, what Jameson means by scarcity is not what Le Guin means by that word, and this has caused confusion among critics. For Jameson, scarcity is the opposite of consumerist abundance, and therefore preferable to it. For Le Guin, scarcity is a problem. It is better to live with scarcity than with consumerist abundance, but it is better still to live under what midcentury anarchists called "postscarcity" conditions. Annares is not a site of desirable scarcity, so much as it is a place where postscarcity conditions might take root, through continuous and nonhierarchical work to end the forms of inequality that inhibit abundance. This other kind of abundance, not consumerist abundance but rather postscarcity abundance, has nothing to do with property or money or the U.S. Instead it is the adjustment, through slow political reorganization, of a people to a planet.

Le Guin, in a 1973 interview, describes *The Dispossessed* as "an anarchist novel" whose anarchism is "simply anti-centralized-state," that is, advocating for "a world with a lot less government, and a decentralized world, and a world without authoritarianism."[30] It is a highly planned lesson in resource management—"How do you distribute goods? Well, if you assume a high technology and a rather small population (which I could do because I was not using the earth, I was using another planet) then things like getting people fed and so on are pretty easy"—but these goods are already in adequate supply, at least on Urras.[31] The challenge on Urras is to rearrange the structures of power, and the practices of making and accumulating, so that the myth of scarcity no longer holds. The challenge on Annares is how to produce enough to go around, or else how to live equitably with very little. If there is a commons in *The Dispossessed*, it is a commons neither of spaceship nor of lifeboat. The commons is instead an ethical and epistemological problem, the challenge of living together. Neither the people of Urras nor the people of Annares have this properly figured out, whence the ambiguity of the subtitle. But between the two planets, Annares is the one that admits to the need for total social and political reorganization. Surely this is world-reduction, in the sense that Jameson means. But rather than reduce Urras to capitalism and Annares to anarchism, *The Dispossessed* more fundamentally reduces Urras to the drive toward institutional reform and Annares to the drive toward a postscarcity revolution. It is in this context that the book develops its theory and technology of communication and communion, not around a reformed idea of message transmission, but instead around the revolutionary ethical condition of communion.

To the extent that the novel is more than a "sermonette" in political science, *The Dispossessed* follows Shevek's efforts on Urras to develop mathematical underpinnings for the ansible. Shevek's research is stalled until he becomes preoccupied with Albert Einstein's writing on relativity: here remembered as a Terran scientist named Ainsetain, author of a "symposium on the theories of Relativity."[32] Two things happen with the introduction of the Terran's book, one textual and one thematic. Textually, *The Dispossessed* immediately takes on the generic character of a future history. Whereas it had not previously been clear how much the novel's "Terra" has to do with the actual Earth, that distant planet now becomes quite real with the introduction of a historical personage. His name has been corrupted by the passing centuries, from Einstein to Ainsetain, but the reader has learned this is a real universe rather than a fantastic one. Thematically, at the same time, the novel provides a methodological basis to the scientific and mathematical discovery of its protagonist. What Shevek derives from the inventions of the fictional character Ainsetain are in fact Le Guin's derivations from the work of the real Einstein. The ansible, the novel's narrative invention, is therefore grounded not just in the internal logic of the book's diegesis, but also in the tested scientific practices of the extradiegetic universe.

In particular, Shevek is drawn to Ainsetain's neglect by peers and his consequent failure in his own time. Shevek too feels like a neglected failure, and so finds inspiration in Ainsetain's story: "He had not succeeded. Even during his lifetime, and for many decades after his death, the physicists of his own world had turned away from his effort and its failure, pursuing the magnificent incoherences of quantum theory with its high technological yields, at last concentrating on the technological mode so exclusively as to arrive at a dead end, a catastrophic failure of imagination." Contemporaries and inheritors had turned away from Ainsetain's theories of general and special relativity, and instead toward an inquiry that was principally technological in both its means and its ends. Shevek's discovery, the book suggests, will derive not from the "failed imagination" of a quantum mechanics driven by the desire for greater "technological yields," but instead from Ainsetain's theorization of "indeterminacy."[33]

Ainsetain had been underappreciated and his theories overlooked for "the magnificent incoherences of quantum theory," Shevek decides, because Ainsetain had found insufficient proof. And yet, Shevek argues to himself, "in the region of the unprovable, or even the disprovable, lay the only chance for breaking out of the circle and going ahead."[34] Like Ainsetain, Shevek has long been engaged in a contrarian theory of temporality. It is widely accepted in Shevek's universe, as in our own, that time moves sequentially and in one direction. Shevek meanwhile has built his considerable and controversial fame on the idea that time only feels sequential, but is more fundamentally a matter of simultaneity. Explaining his theory to another character, he employs the metaphor of a book: "Well, we think that time 'passes,' flows past us, but what if it is we who move forward, from past to future, always discovering the new? It would be a little like reading a book, you see. The book is all there, all at once, between its covers. But if you want to read the story and understand it, you must begin with the first page, and go forward, always in order."[35] This is Shevek's own unprovable theory. Time is not the progress of instants. It is not a story of being, a steady rhythm that humans find a way to measure. It is instead a closed book in which everything happens at once, but through which each being must pass. Just as Ainsetain had stuck to his disprovable theory, Shevek concludes, so must he proceed with his theory as if it were true. "By simply assuming the validity of real coexistence," that is, of simultaneity, "he was left free to use the lovely geometries of relativity."[36] In opting for simultaneity over sequence, or rather by seeing sequence as simultaneity's mere form of appearance, Shevek opts for communion over communication. Communication can only be conducted in apparent sequence, in the passage of a moment, no matter how brief, between the sending and the receipt of a message. Communion, by contrast, is real coexistence. It means living together. It is not the passage but the sharing of a moment. Moreover, *The Dispossessed* is not just a story whose key metaphor is a book. It is also a book that becomes, with this metaphor, a theory of the book. A book, according to this book, is a device that makes coexistence possible. It does not communicate. Shevek's theory is not just a philosophical physics of sequence and simultaneity, but is more importantly a narratology of diachrony and synchrony. The problem that confronts Shevek is not only the problem that confronts Albert Einstein but also the problem that confronts Mikhail Bakhtin.

The literary chronotope, writes Bakhtin, is a "vertical world" in which "everything must be perceived as being within a single time, that is, in the synchrony of a single moment; one must see this entire world . . . under conditions of pure simultaneity . . . to replace all temporal and historical divisions and linkages with purely interpretive, extratemporal, and hierarchized ones."[37] So it is with Le Guin's book, and with the metaphor-of-the-book that it contains. Sequence is nothing but a feeling. Chronographic technologies do not measure external phenomena indifferently. Rather they regiment sequence, in an activity that is both hierarchical and interpretive. To cohabit in the world with others, it then follows, is not to strip away the technology as if moving toward an unregimented form of sequence. It is instead to see sequence as a mere secondary effect of a dehierarchized simultaneity that occurs before, or without, any act of interpretation. A book does not communicate. Instead it facilitates communion. The same is true, ultimately, of the ansible itself, which makes possible a sharing of words and ideas over long distances—less through the sequential transmission of messages and more through an act of becoming joined with others: "thus approached, successivity and presence offered no antithesis at all."[38]

Communion and real coexistence—these are the principles of the ansible and of the book (that is, of books: the book that Shevek describes in *The Dispossessed* and the book that *The Dispossessed* itself is). They are also political principles, incompatible with the capitalism of Urras. They might eventually make sense on Annares, but even on Annares they do not yet make sense. What Shevek cannot understand about Urras, and what his Annaresti upbringing leaves him unprepared to comprehend, is the fact of unequal distribution. And yet, contra Jameson, Le Guin does not stage this fact as a problem of adequate or inadequate quantities of available resources. Instead, she stages it as a problem of mutual obligation. As Shevek tells his valet Efor, Annares is a place where "we have been hungry" and where "I knew a woman then who killed her baby, because she had no milk, and there was nothing else, nothing else to give it"; yet in spite of that tragedy, "nobody goes hungry while another eats" (285). In short, Annares is preferable to Urras because no one there gets rich on the back of anybody else. But where starvation still happens, as it does on both Urras and Annares, there is no unambiguous utopia.

The path of the novel's utopian vision is toward a future version of Annares in which the problem of relative scarcity or relative abundance would fall away. In its place would be a political system that begins, but does not end, with the principles of communion, simultaneity, synchrony, and real coexistence. "Nobody goes hungry while another eats" is a class principle that becomes a principle of literary temporality and experimental physics, and it is on such a principle that something better than Urras, and even better than Annares, may be built. It has not yet been built, and yet it may be, as Shevek returns home with his theory of real coexistence at the novel's close. Shevek explains the possible eventual outcome of this theory to a Terran ambassador: "Space travel, you see, without traversal of space or lapse of time. They may arrive at it yet; not from my equations, I think. But they can make the ansible, with my equations, if they want it. Men cannot leap the great gaps, but ideas can."[39] Likewise, while the novel does not imagine concrete practices by which to overcome the opposition between scarcity and abundance, it nevertheless employs imaginative narrative as a way to "leap the great gap" that separates hunger from its opposite, toward the idea of another kind of communion.

The End of Meaningful Communication?

Like Le Guin, Samuel R. Delany develops a theory of communion during the Long Seventies. Delany too uses the word *communion* to describe a way of living together, modeled by literature, apart from cosmotechnics. As much a writer of theory as of fiction, Delany signals the notion of communion in his 1976 novel *Triton: An Ambiguous Heterotopia*, and develops it in his 1977 introduction to a republication of Theodore Sturgeon's *The Cosmic Rape* (1958). In Sturgeon, Delany identifies a structure of togetherness, of community or *communitas*, that can approach a form of interaction that is even more intimate:

> This order of *communitas* always on the verge of communion expresses an inchoate need in the American psyche. . . . Looking at the range and power of this communion as it is presented again and again throughout Sturgeon's work, certainly I see love as one of its most important forms. Yet what has always struck me . . . is how much larger than love—love in any form I could recognize—this communion is always turning out to be. It is almost always moving toward the larger-than-life, the cosmic, the mystical. In a number of places in Sturgeon's work it becomes one with evolution itself.[40]

Le Guin had construed communion as real coexistence, interplanetary simultaneity, life lived in the time of the book rather than the time of telecommunication. Delany construes as communion a way of being together that exceeds all known ways to be together. Community approaches communion, but does not meet it. Love is communion, but communion is not only love. With all the closeness of love, and perhaps of sex, communion operates at the scale of the cosmos.

Most important in this tendency toward cosmic scale, Delany describes communion as "always moving." It is not a pool of shared resources. It is susceptible neither to scarcity nor to its opposite. It is instead a motile and expansive procedure of affect and attachment, whereby the universe is brought together through the peace and pleasure of something that is larger than love. In Delany's remark that communion can "become one with evolution itself," it is fair to hear a critique of Buckminster Fuller. For Fuller, humanity was not properly human until it invented its "total communications system," the network. For Delany, however considerable the impact made on the species by satellites and computers, this impact is not evolution. Evolution is cosmic intimacy. It is community beyond community, on the verge of something that is not community, communion.

The question remains as to whether this communion, in Delany, is the same as "real coexistence" in Le Guin; and whether it is, as communion is for Le Guin, "opposed to all the hollow forms of communication." As it happens, Delany's 1976 novel is in part a critique of Le Guin's novel of two years before. Even though its subtitle was added after the novel was mostly complete, *Triton: An Ambiguous Heterotopia* must yet be read as an implicit response to *The Dispossessed: An Ambiguous Utopia*. The main point of contact between the two books is in the shift from utopia to heterotopia, which Delany makes explicit in a chapter epigraph from Michel Foucault. Whereas utopias can be said to "permit fables" of verbal comparison between good and bad places (*topoi*), writes Foucault in the passage from *The Order of Things* that Delany quotes there, a heterotopia is neither a good place nor a bad place, but is another place entirely. A heterotopia is a different place, and a place of difference, that can "desiccate speech, stop words in their tracks, contest the very possibility of grammar at its source."[41] *The Dispossessed* compared two places, Urras to Annares, in terms that were philosophically ambiguous but narratively quite lucid. With world-reduction as its primary tactic, it is never unclear what is happening to Shevek. It is never unclear that the worlds exist in a fable of comparison, even when a reader is unsure which political principles inhere in which world. By contrast, *Triton* is neither politically nor expressively reducible, but is instead an experiment in the "desiccated speech" of another place, where words "stop in their tracks."

Avowedly heterotopic in this way, *Triton* is a formalist critique of the ideal of linguistic communication, as well as a critique within language of any ideal form of politics. *Triton* tells a story of a community on the Neptunian moon of the title, where sexual mores and familial norms have been wholly rejected; and it tells at the same time of a character named Bron, whose tendencies toward misogyny and homophobia have left him excluded from that community. Bron comes to crisis over his reactionary impulses, then as a last-ditch effort to resolve this crisis, transforms chemically and surgically into a woman. Yet even as a woman, Bron clings to sexual and gender hang-ups that keep her excluded. Early in the book, Bron is introduced to the lives of the street gangs, political cadres, and theatrical troupes that live in the "unlicensed sector" of the city. This is an area in which social deviation has been legalized and localized, leading to stability in the overall social structure. It is the another-place of the novel's subtitle, and it is where Bron finds love with a radical dramaturge and performance artist named "the Spike."

One of these roaming gangs, the Spike explains, is called the Rampant Order of Dumb Beasts. It is a "neo-Thomist sect" that has committed to "putting an end to meaningless communication. Or is it meaningful . . . ?

I can never remember."[42] Shortly after this first meeting, having witnessed one of the Spike's performances, Bron walks away deep in thought: "Were the Dumb Beasts, he wondered suddenly, also part of the charade? . . . Meaningless communication? Meaningful . . . ? Which one had she said?"[43] It is a joke about communication. Combining elements of spatial and linguistic alterity, the same joke is evoked a few times through the first part of the novel, providing it a rhythm and transforming with each iteration. Sometime later, the Spike introduces Bron to a member of the Beasts named Fred, about whom Bron then complains: "Your friend doesn't seem very communicative." Unperturbed, the Spike explains: "The whole problem, I suppose, is that any time some piece of communication strikes poor Fred, or any of the remaining Beasts, for that matter, as possibly meaningful—or is it meaningless? It's been explained to me a dozen times and I still can't get it right—anyway, his religious convictions say he has to either stop it or—barring that—refuse to be a party to it."[44] The Spike effectively explains the challenge of Fred's sect. Yet she does so without knowing, or without admitting, from which kind of communication the Dumb Beasts abstain: meaningful or meaningless. The Spike's own utterance is itself meaningful. But, thick with em-dashes and split down the middle by a question mark, it approaches meaninglessness, not only because it switches directions throughout, but also because it revolves around an indecision, a question mark. Communication, it seems, can be meaningless or meaningful, but it does not matter which. What matters to the Dumb Beasts, and to the Spike who respects their struggle, is that some communications, or some whole idea of communication, must be stopped or refused. The effect is communion, or something near to it. Like so many of the familial and parafamilial groups in this heterotopia, the Dumb Beasts provide one another with a mystical kind of love, at least until they disband to go join other cadres, other familial and parafamilial communities that verge on communion.

When the Spike eventually dumps Bron, deservedly, she does so in a letter sent to Triton from earth. Employing a "very old voice-scripter" (a technology resembling the voice-to-text feature of an iPhone), the letter includes not only punctuation marks and paragraph breaks, but also the Spike's verbal instructions as she tells her broken device what to type: "Bron, and then I guess you better put a colon no a dash—the world is a small place italicize is. And the moons are even smaller."[45] Evoking the image of a shrunken world, that central cosmotechnic metaphor, the Spike writes a letter that is again both too meaningful and not meaningful enough. Too meaningful because it includes both the proper punctuation (a dash) and also the Spike's indecision about which punctuation to use (the words "a colon no a dash").

Not meaningful enough because the technical error has produced a nearly nonsensical prose:

> What do I want to explain?
> That I don't like the type of person you are. Or that the type of person I am won't like you. Or just: I italics don't. Do I have the colon in there? Yes.[46]

The Spike is deliberate in her prose, too deliberate; so deliberate that she nearly ceases to make sense when she doubles back to check on colon placement. She goes on, brutally but not inaccurately:

> You do italics adhere to some kind of code of good manners, proper behavior, or the right thing to do, and yet you are so emotionally lazy that you are incapable of implementing the only valid reason that any such code ever came about: to put people at ease, to make them feel better, to promote social communion.[47]

Here, in the voice of the novel's most ethical character, is Delany's version of communion. Yes, communion is intimacy that expands beyond community to encompass the whole of evolution and the cosmos. Yet this intimacy hinges neither on clear communication nor on any clear distinction between meaningless and meaningful communication, nor on any easily broken technology of communication, like this "very old voice-scripter." Instead, it hinges on the clichés and artifices of basic social interaction ("proper behavior") and minimal ethics ("the right thing to do").

Describing *Triton* to the fan magazine *Algol* in 1976, Delany explained that the book's original main title—*Trouble on Triton*—was meant to feel instantly familiar to readers of science fiction: "I wanted a title that would evoke a sense of 'Haven't I heard of this before?' and the title would sort of slip between your fingers before you actually grasped it. . . . It is a book about people trying to live their lives by clichés."[48] It is this last part that the Spike means when she tells Bron that the latter is "incapable of implementing the only valid reason" for social and behavioral codes. What the Spike knows, but Bron does not know, is that clichés exist only as fungible modes of entry into social communion. Bron's error is an adherence to codes for their own sake. The novel's other characters—the Spike, Fred, the Rampant Order of Dumb Beasts, and everybody else—use codes only so they may find ways to coexist with others. Then they let those codes "sort of slip between their fingers," like so many useless clichés or vaguely familiar titles, before they can cohere into any superficial but more permanent form. Clichés are a routine, a rhythm to social space, the reality of simultaneous coexistence, and they communicate nothing, but facilitate communion nevertheless. This is what Lauren Berlant frequently means by "cliché." In writing about Bette Davis's character in the film *Now Voyager* (dir. Irving Rapper, 1942), Berlant insists that clichés are not real ideals, but they do point a way forward toward autonomy under the constraints of contemporary life: "All the women need to do is to pronounce the phrases and make that journey toward the something else not quite visible, yet. . . . All she ever wanted is the right to live her own clichés. . . . You might say this is what freedom is in liberal culture."[49] Berlant's something-else-not-quite-visible is communion. The Spike tells Bron that this is what clichés are for: not for establishing hard and fast codes of ethics, but rather for enabling the degree of freedom that is possible, for bringing community a little closer to communion, and then for slipping away.

Unsurprisingly, considering the provenance of *Triton*'s subtitle, this critique of communication likely owes something to Foucault. Communication is not a synonym for language or speech, for Foucault, and it is certainly not the name of the fulfilled evolutionary promise of the species. In "The Discourse on Language," his 1970 inaugural lecture for the Collège de France, Foucault explains that communication is a historically specific idea of language, but not language itself:

> To the monopolistic, secret knowledge of oriental tyranny, Europe opposed the universal communication of knowledge and the infinitely free exchange of discourse. This notion does not, in fact, stand up to close examination. Exchange and communication are positive forces at play within complex, but restrictive systems: it is probable that they cannot operate independently of these.[50]

Foucault here finds communication to be a discursive construction misapplied to practices outside its constitutive constraints. The Eurocentric model of communication, for Foucault, develops out of its seeming difference from non-European models. It names the ideally unregulated circulation of information that would undermine any tyrant who would keep its secrets for his own. Yet communication never really was unregulated, but was always conducted within the defined parameters of particular places and times.

These parameters are what the Spike calls codes, and what Delany and Berlant call cliché. Within them there takes place either a familiar ritual of communication, facilitated by technologies of ever-greater reach and clarity, or else a radical move toward communion. The ritual of communication holds to the codes, pressing them toward more and more efficient conduct within the parameters of the "restrictive system." Communion arises in another kind of language, a heterotopic kind, breaking away from systems and letting go of codes, against communication meaningless or meaningful, desiccating speech and stopping words in their tracks. Communication is not the "free exchange of discourse" at all, but is something else entirely: the rituals of speech that might be conducted otherwise. "Ritual defines the qualifications of the speaker," writes Foucault, just as "it lays down gestures to be made, behavior, circumstances and the whole range of signs that must accompany discourse."[51] Communication is cliché, and includes the practices of language that may occur within the bounds of accepted ritual. Outside of those bounds, language must operate without communicating anything, but that does not mean it stops being language.

The linguist William D. Seidensticker, in the year after Foucault's lecture, writes: "Philosophers who dabble in linguistics and linguists who wax philosophical often seem irresistibly drawn to the view that language . . . is a system of communication."[52] These philosophical linguists and linguist-philosophers, in Seidensticker's ventriloquy of them, give ironic voice to perspectives that have since become commonplace. With a nod to McLuhan, Seidensticker begins:

> "Language is a medium" is the message. Our most ordinary conversations are communications, exchanges of coded messages. We the users of "natural language" simply don't realize what talented communications experts we *really* are. All of the world is a network and we are the senders and receivers. How to succeed at cryptanalysis without knowing it! I find this view of language deeply confusing.[53]

Seidensticker is confused because he sees "communication" as a word formerly used most often to describe the kind of message transmission that takes place outside of speech, in lieu of speech, in emergencies, in silence, or across distances; but that is now expected to define all language, whether exceptional or unexceptional. Communication, in other words, was a fashionable idea that came to be misapplied to dissimilar matters of language. In Seidensticker's indictment of language-as-improvised-cryptanalysis, one might be tempted to hear a rejection of the semiotic view of language as a system of interpretable signs, or a rejection of the study of ideological speech. But really his irony is aimed at the ideology and fiction of language as communication. His target is the idea that what happens in common discourse (conversation, news, rumor, art) is the same thing that happens when a signal is sent through a cable or from a satellite to an antenna. When Seidensticker isolates the perspective from which "all of the world is a network and we are the senders and receivers," he is insisting that most words are local rather than global, and that not every relation is networked, not every speech act a message.

Unfree Flow

Internal to the idea of communion, an idea that emanates from Le Guin and Delany to resonate with Seidensticker and Foucault and Berlant, there is posed a challenge to the very idea of networks: if communication is not inclusive of every speech act, then all the world is not a network, and humans are not (or cannot merely be) senders and receivers. The cogency of this theoretical position does little to dismantle the industrial and ideological practices that it opposes, for which networked communication is an unquestioned fact, and the scarcity-bound metaphor of Spaceship Earth continues to hold. The goal of the more dominant practices is the "free flow of information," a phrase that may not make any sense in the critical theory of language, or in its science-fictional variations, but whose implications are entrenched well beyond theory, in the conduct of internationalist institutions. The doctrine of free flow, as it is often called, finds its most explicit form in the constitution of the United Nations Educational, Scientific, and Cultural Organization. In fact, the first article of the UNESCO constitution states that the mission of the organization is to "contribute to peace and security" by a number of means, including "the work of advancing the mutual knowledge and understanding of peoples, through all means of mass communication . . . to promote the free flow of ideas by word and image."[54] This constitution was written in 1945, nearly two decades before the launch of Telstar 1, and still a year before the establishment of the first television network, NBC, in 1946. Yet even at its outset, free flow was imagined as a way of using mass communications toward cross-cultural understanding and planetary security. The metaphor of the world-as-network, alive already in the UNESCO constitution, is the very principle of cosmotechnics. And because it made no provision for the wealth and class disparity involved in the production and distribution of information, the free-flow doctrine soon became the target of anticolonial critics of cultural imperialism.

Herbert I. Schiller, the U.S. theorist most associated with the coinage and critique of "cultural imperialism," was also the most visible public critic of free flow. Arguing that the doctrine requires and enforces an illusion of parity between and among nations with unequal resources, Schiller concludes that costs are very high for the world's less wealthy peoples. For Schiller, in 1974,

> what people believe, what they aspire to, and what moves them to act or not to act constitute an essential part of the community's living pattern. To permit this pattern to be subjected to external influence and control would seem unthinkable; yet, till recently, this has been the rule, not the exception. The free flow of information, reinforced by economic power, has led to a worldwide situation in which the cultural autonomy of many (if not most) nations is increasingly subordinated to the communications outputs and perspectives of a few powerful, market dominated economies.[55]

What Le Guin calls "communion" is what Schiller calls "cultural autonomy." What she calls the "hollow forms" of communication are the devices that house the illusory freedom that Schiller indicts. A real unfreedom and a failure of communion result from the doctrine of free flow, especially for those who live outside the centers of culture and capital. On the receiving end of cultural production, Schiller argues, the living pattern is disrupted; moreover, such disruption is "the rule, not the exception."

Cultural imperialism is in this way not merely enabled by new technologies. It is enabled by an idea, the doctrine of free flow, the formula through which new technologies are legitimated and commandeered. As Belgian theorist Armand Mattelart put it in 1973: "Lately the electronic and the air and space industries have ceased to live electronically, and have made their entrance into the contingency of politics";[56] and moreover, "no operation better reveals the phantasmagoria of the technological Messianism of imperialism, as does a cold analysis of the power relations within its economic expansion."[57] For Mattelart, working at the same time and in conversation with Schiller, communication has taken on a material role quite different from the sending and receiving of messages. More important than the function of transmission or the words transmitted, and more important than its composition out of circuits or satellites, the electronics industry is a global extension of power, a capitalist machine. It is an ideological formation, "technological Messianism," that promises to change the world suddenly and totally, by converting that world into a global information marketplace. This promise remains in force, as Mattelart writes on returning to the topic in 1983: "The arrival of Reagan has drawn together the defenders of the free flow of information under the banner of neo-liberalism."[58]

This is the foundation of the metaphor of the network. Following Seidensticker, language does not in fact transform its speakers into communicative senders and receivers. Following Foucault and Berlant and Delany, communication takes place only within the bounds of its ritual, and is therefore antithetical to any flow of information that is really free. Following Schiller, the doctrine of free flow enables only a unidirectional imperialism of information, disabling any local efforts at community formation. Following Mattelart, this anticommunalism is part and parcel of U.S. capitalist expansion, especially as regards cultural and media policy. The network of communications technology, it turns out, is even more importantly political and economic than it is communicative or technological. How then to proceed? If the network metaphor has provided the most convincing contemporary image of a world in process, but that metaphor is fundamentally bound up with capital and empire, what alternative metaphors might be possible? If planetary communion is not to be achieved by overcoming the barriers to communication, but instead by overcoming Le Guin's "hollow forms of communication," then how does it start?

Building a Common World

On Christmas Day in 1968, as the Apollo 8 shuttle began its return to earth following the first human orbit of the moon, the *New York Times* published an appreciation in prose, by the poet Archibald MacLeish. On the front page, below the gripping news of the returning astronauts, MacLeish gushed: "To see the earth as it truly is, small and blue and beautiful in that eternal silence where it floats, is to see ourselves as riders on the earth together, brothers on that bright loveliness in the eternal cold—brothers who know now that they are truly brothers."[59] MacLeish's sentimental lines refer at once to three aspects of the political moment: one, a technological vision, a planet transformed by humanity's new capacity to see it "as it truly is"; two, a geopolitical vision, a world made eternally cold by a Cold War; and three, a vision of cosmotechnics that would seek to utilize technology to transcend geopolitics. It is this last vision, of humanity "as riders on the earth together," that fuels hope in networks and planetary spaceships.

At last, is it even possible to be with each other in the world? Who is this "we" that ride the earth together? Philosopher of technology Jean-Pierre Dupuy posed the problem in this way in Milwaukee in 1977: "The belief that men will not really be able to communicate with each other until they are freed from the sway of things is untenable. Relationships between people are defined in and through the process of making things together, building a common world."[60] Building and making things, in short, are opposed to the free flow of information. They have nothing to do with the fictions of a lifeboat or spaceship that is shared and resource-poor. They have nothing to do with the geomilitary employment or capitalist control of networks. To be sure, literary world-making is a narrative and aesthetic procedure, so it does not escape the distributive pathways of capital. Yet even from the confines of its distributive pathways, world-making issues a competing model for communion in the present. Dupuy concluded: "Action, like speech, is naturally transient and cannot remain in the memory of men unless actualized by poets and historians. Hence men cannot communicate unless they have a common world which unites them, but also separates them, just as a table brings people together insofar as it stands between them."[61] This is a model for a communion founded in language, and one that is wholly different from the communicative network. It is impossible to let go of the metaphors of communication altogether, but it is certainly possible to recognize that political deliberation and ethical togetherness are ill served by metaphors of free flow and imperatives toward unobstructed message transmission. Opposed to technological Messianism, then, is the image of a device far simpler than Telstar 1 or Arpanet: a table, whose function is both to obstruct and to unify the rituals of discourse, facilitating real coexistence, tinkering in the mechanics of cliché, verging on communion.

CHAPTER 3

Cyberculture

This is our chance—"our" meaning all of humanity, not only a few who are exceptionally fortunate—to . . . do our human work, to live human lives, to devote ourselves . . . to poetry and politics.

—Alice Mary Hilton, at the Conference on the Cybercultural Revolution (1964)

At least since the mid-twentieth century, universalist hopes for global peace and collective productivity have been linked to the devices of accelerated travel, labor, and telecommunication. This is cosmotechnics. In the United States in the 1960s, the link appears forged in consensus, and expressed most acutely in the dedication of the 1964 New York World's Fair. The fair's theme was "Peace through Understanding," and its dedication was to "Man's Achievement on a Shrinking Globe in an Expanding Universe." The divergent ideals of peace, species exceptionalism, telecommunications, and space travel carry no natural or necessary relation to one another, but here they fold tidily into a unified ideological formation—one in which technological change might lead directly to cross-cultural "understanding" and thence to peace. This is a humanist pastiche. From it diverges a contrary project, differently humanist, in literature and politics. That contrary project, then called cyberculture, is the subject of this chapter. At once aesthetic and political, cyberculture aims to disturb techno-centric and cosmotechnic thinking through a pacifist and antiracist critique of the myth of resource scarcity. Threaded through a range of literary and political texts from the early years of the Long Seventies, cyberculture offers a vocabulary in excess of available language. Dismantling is Luddism, a way to break, study, or slowly relinquish machines; and dismantling is communion, a critical practice of togetherness aimed at unsettling the presumptions of technology, scarcity, and "free" communication. Dismantling is also cyberculture: an ethical regimen gathering humans toward shared commitments in their machine environment; enabling a new devotion, following Alice Mary Hilton, to poetry and politics.

A Cyber Nation

On the Fourth of July in 1964, two apparently unrelated reports appeared in the Talk of the Town section of *The New Yorker* magazine. One report covered a meeting of the executive committee of the World's Fair in Flushing, Queens. Showing the early cracks in the architecture of cosmotechnics, the story covered the controversy over a mural in the Jordan Pavilion, entitled "Mural of a Refugee." Explicitly a defense of Palestinian rights, the mural was a large black-and-white image of a mother and child, beside a poem that was printed on the wall, beginning: "Before you go, / Have you a minute to spare, / To hear a word on Palestine / And perhaps to help us right a wrong?"[1] Under discussion by the executive committee, in fact, was whether a discussion could even take place in regard to this mural. To members of the committee who wanted the mural removed for its ostensible offense to Israel, the fair's president, Robert Moses, replied that the item could not be removed—or publicly discussed—until related cases were resolved in the state's supreme court. As the possibility of dispute was itself being disputed, the unnamed *New Yorker* contributor delighted in this notable contradiction: How could a fair with the theme "Peace through Understanding" and a dedication to "Man's Achievement on a Shrinking Globe in an Expanding Universe" be so marred by conflict and clamor? After a frenzied climax in which "Mr. Moses kept banging away" with his gavel, the story ends on a wry note: "After that, everyone watched a movie called *Great Fair, Great Fun.*"[2] When its celebration of communication technology is paired with its failures of communication, this World's Fair serves as a backdrop for subsequent critiques of those same regimes: communication, technology, and globalism.

The second item in *The New Yorker* that Fourth of July was a report on two events: the first assembly of the Society of Women Engineers, at the United Engineering Center; and the Conference on the Cybercultural Revolution, in the Terrace Ballroom of the Roosevelt Hotel. Organized by Alice Mary Hilton, the latter event was described in typical Talk of the Town style, as if it were a tea party whose star was not Dr. Hilton, logician and computer scientist, but rather "Miss Alice Mary Hilton, who was wearing a dark-green dress."[3] Unmentioned in *The New Yorker* is the participation of luminaries from far outside the study of technology, from the philosopher Hannah

Arendt to the racial justice advocates James Boggs and Grace Lee Boggs. Of concern to Hilton, the Boggses, Arendt, and others were the new forms of workplace automation that Donald N. Michael, another technologist in attendance, had called *cybernation*. Hilton defined this term for her interviewer in *The New Yorker*:

> With automation, you need people to run the machines, and you can create more jobs by increasing production. But cybernated machines don't even have control panels; control panels are for people, not machines. Cybernated machines run themselves, and people are superfluous.[4]

Hilton did not object to cybernation, which she regarded as a stage in human progress but also as a tipping point, where things could go either way, toward more exploitation or toward less. Were cybernation to be done right, Hilton argued, there would be fewer exploited workers. This is one aspect, but for Hilton the most central, of a wholesale shift in human knowledge and potential. Life in an economy of scarcity, where accumulation is a zero-sum game that can benefit only some, at the expense of others, would be replaced by an economy of abundance. But cybernation, the conferees agreed, would result in more resources than necessary, and therefore less labor to be performed. With less pain and more resources, humanity would enjoy new prosperity and leisure.

For Hilton, indeed, if there were fewer humans doing less exploitative labor, then so much the better, as long as these humans were provided the means to thrive. Moreover, for Hilton, cybernation would mean a transition from *labor* to *work* in the way that Arendt had distinguished these terms. In an interview with Studs Terkel the next year, Hilton explains: "It's so important to differentiate between work and labor. As Hannah Arendt points out . . . work has always had the connotation of something creative, of something you do for the sake of doing it. Labor has always had the connotation of coercion, of pain and toil."[5] The arrival of these "thinking machines," as Hilton and others had begun to call them, would entail an expansion of those spheres of life that could properly be called "work" rather than "labor." Less time in the field or the factory, for example, would mean the growth of domestic activities and activist practices, as well as art, science, games, and other pleasurable creative activities on which humans would be newly liberated to spend their time.

Hilton's adoption of Arendt's framework is only partial. It concerns Arendt's distinction between work and labor, and excludes the third term— *action*—in what was in fact a well-known triad. In Arendt's *The Human*

Condition, labor is the natural and "biological process of the human body," while work is not natural at all, but is instead what "provides an 'artificial' world of things." Work is preferable to labor in Arendt, but action takes priority over both. Action is the free and unmediated congress of human beings, and not only creates "things," as work does, but also "creates the condition for remembrance, that is, for history."[6] Then not only subordinating work and labor to action, Arendt also subordinates the whole triad, to which she gives the name *vita activa*, to another kind of life to which labor, work, and action might give way: the life of the mind, the *vita contemplativa*. As Arendt clarifies in 1964, later in the year of Hilton's interview in *The New Yorker*, "the *vita activa* was always defined from the viewpoint of contemplation."[7] For all that separates work from labor, and for all that makes action superior to either work or labor, when "compared with the absolute quiet of contemplation, all sorts of human activity appeared to be similar insofar as they were characterized by unquiet."[8] Hilton's appeal to Arendt is quite limited indeed, then, inasmuch as Hilton leaves out both the trump card of action and also the ultimate goal of contemplation. Is this omission deliberate or accidental? A theoretical intervention or an error derived from a casual reading of *The Human Condition*? Intervention seems no less likely than accident, as Hilton's analysis never promises any free discourse, unregulated contemplation, or an ideal of immediacy. At stake instead is a potential shift from exploited labor to creative work, and the consequent growth of leisure time in an economy of abundance. Action and contemplation would only be secondary effects of this shift.

In organizing the conference, then, Hilton opted for a word that would capture the risk and possibility that inhere in the idea of abundance, as well as the breadth of its mediated forms of expression, forms she had named the previous year: cyberculture. For many of the conferees, cyberculture named a complex intersection of cybernation with the similarly changing technologies of militarism and institutional racism. For all of them, liberals and leftists, extreme forms of automation must spell the end of a dream of full employment (as jobs were and would be disappearing from every field) and the renewal of another dream, the end of exploitation. This is where the cyberculture conference departs from what is now known as accelerationism.[9] At the cyberculture conference but unrecorded by Talk of the Town, Hilton consolidated her technicist's humanism in the discussion that followed Arendt's lecture. She told those assembled: "This is our chance—'our' meaning all of humanity, not only a few who are exceptionally fortunate— to . . . do our human work, to live human lives, to devote ourselves . . . to poetry and politics."[10]

Hilton reiterated these priorities in a 1968 issue of the journal *Improving College and University Teaching,* where, tucked between articles on pedagogy, there appears a poem called "Hilton's Law." Using language more polemical than poetic, as perhaps suits her nonliterary profession, Hilton lays out a code of ethics in two stanzas. In the first stanza, Hilton lists social and cultural roles that are valuable to the democratic lives of communities. In the second, she offers an opposing list of social and cultural roles that put democracy and community at risk. The first line is "A society building the City of Man* must cherish:"—and its asterisk directs readers to a footnote at the bottom of the page: "per Walt Whitman; also called the City of God, per St. Augustine; it may perhaps be called the City of Human Love or of Politics and Poetry by some future cybercultural poet."[11] This footnote is significant for several reasons, not least because it signals a debt to Whitman—a debt that is made both thematically and formally evident in the lines that follow. Thematically, Hilton is concerned with the social and poetic preconditions for radical democracy. Formally, her poem imitates the lists in *Leaves of Grass* that James Perrin Warren has called Whitman's "clausal catalogues."[12] Several indented lines make up Hilton's own clausal catalogue, so that she may tell us precisely what it is that "a society building the City of Man [or City of God, or City of Human Love, or City of Politics and Poetry] must cherish":

the skeptic, but not without examination,
     lest he be merely a cynic;
the dissenter, but not without gentle questioning,
     lest he be merely an oppositionist;
the iconoclast, but not without honest doubt,
     lest he be merely a destroyer of beauty and dreams[13]

Hilton's Whitmanian or Augustinian city is to be built on revolutionary dissent, rather than on such consent as might be presumed by the social contract. Moreover, her city is to be founded not on the sacralization of images but instead from the process of iconoclastic dismantling.

In spite of what might be assumed about the *cyber* in "cybercultural poet," Hilton's revolutionary poet needs never to have had any direct contact with computers. Rather, he or she must only have lived in a world where human associations have been definitively altered by the introduction of such machines. In Hilton's reading of Norbert Wiener, the science of cybernetics is not properly a science of computers at all, and neither is it a science of automation or communicative networks or robotics. Rather, she writes, cybernetics is "the science of relationships," such that "a cybercultural society is, therefore, an integrated society that affords human beings the opportunity to live human—i.e., purposeful and civilized—lives."[14] To Studs Terkel, Hilton argues, "cybernetics is confused with the theory of machines, which it is not—that's just one of the applications"; instead, cybernetics should name "the idea that relationships like this can be expressed accurately and mathematically."[15] Cyber*culture*, in turn, describes a culture whose constituting relationships can be understood through observation and analysis.[16] For Hilton, such a society will be enabled by computation and automation, to the extent that computation and automation can create conditions of resource abundance and meaningful work. Once such conditions are achieved, there begins the hard work of living together.

The cybercultural poet does not sing in celebration of the computers, but only of a world in which computers exist, and where the living of "purposeful and civilized lives" has become possible through politics and poetry. Cybercultural society is a world of relationships that permit society to thrive. New technologies have transformed lived conditions in which people might relate, but they are secondary to the task of human connection that will follow. There is no suggestion by Hilton, as there is in contemporaneous visions, that the world has gotten smaller. Moreover, in cybercultural society, there are not only persons whom "a society building the City of Man must cherish." There are also persons whom "a society that wants to avert its own decay and demise must suspect." Of these latter, Hilton lists the conformist, the booster, the reactionary, and the patriot, each of whom may in fact be "an optimist where optimism is reasonable," but is probably just a cynic or opportunist.[17] Cynical conformists and boosters, reactionaries and patriots, the poem argues, would only thwart the City of Human Love or the City of Poetry and Politics, which the earnest skeptic, dissenter, and iconoclast might bring into being.

A Vision of Democratic Life

The actual coinage of the word *cyberculture* occurs in Hilton's 1963 book, *Logic, Computing Machines, and Automation*, which combines social theory with computer science.[18] When first used, the word could not have named what it now appears to name: a digitally immersed world of total mobility, instant communication, and universal information access. Yet in its initial use and, more important, in its early inheritance, something new did emerge: not a new technology and not a new way of using old technology, but rather an awareness that technologies of communication and computation share a history and a fate with other technologies. In Hilton's coinage, cyberculture is not a trend that takes root, becomes normalized, and later becomes susceptible to critique. Rather, to name cyberculture is already to have begun a critique whose objects are the disparate new technologies and the transforming conditions of life. Cyberculture involves a mutual dependence among technologies that sustain and regulate racism, militarism, and poverty. The act of naming initiates a protest against the militarized and racialized conditions of global labor in the recent past. Not a protest against thinking machines, this naming does nevertheless interrupt the advancing consensus that such machines had shrunk the globe.

Such critique has roots in the writings of Wiener, as well as Arendt and Marx, and extends its branches (as I show in the next chapter) through writing as diverse as science fiction, pacifist poetry, and the polemic of feminist and racial justice. Rather than tell a story about technologies that affect and transform the world, as if autonomously, cyberculture instead tells a story of humans who make technology for human purposes, whether through desire to share and remember, or through desire to exploit and do violence. With the passage of critical paradigms, the voices of politics and poetry become more difficult to hear. They speak quietly, and rarely to each other; but they do remain audible as a protest against the technological violence of war and industry, racialization and class exploitation, in the decades that followed.

The critics of cyberculture and the advocates of cosmotechnics are as distinct from one another as the New Left is distinct from the New Communalists. This latter distinction is drawn by the media historian Fred Turner, for whom the public advocates of new technology and the public advocates of sociopolitical change are different groups motivated toward different ends. The New Left, under the sign of organizations like Students for a Democratic Society (SDS) and the Student Nonviolent Coordinating Committee (SNCC), can fairly be described as having rejected the values of their parents, particularly racism and militarism but also anti-communism. The younger New Communalists, however, wanted something else entirely. Turner writes that "the communards of the back-to-the-land movement often embraced the collaborative social practices, the celebration of technology, and the cybernetic rhetoric of mainstream military-industrial-academic research."[19] The epigrammatic example of this celebration is Brand's Whole Earth, for Turner, because of its efforts to cultivate life unconstrained by the shackles of history, prescriptive ethics, or sexual and psychological norms. "For the New Communalists," writes Turner, "the key to social change was not politics, but mind."[20] From the standpoint of the 1960s, Turner's account explains much, including the fact that many of the counterculture youth grew up to become libertarians and capitalists.

However, in order to explain cyberculture, an activist tendency that continues into the next decade, a supplementary account becomes necessary. Hilton and her colleagues at the cyberculture conference were New Left, not New Communalists: not only because their membership includes leaders of the civil-rights and peace movements, but also because of their vision for the world. While the New Communalists entered the Long Seventies by bringing their ideals of a one-town-world to Silicon Valley, the critique of cyberculture remains a pursuit of politics and poetry, informing activist and literary practices for a decade.

If one were to look up "cyberculture" in the *Oxford English Dictionary*, one would find that its first appearance is a line from Hilton's 1963 foreword: "In the era of cyberculture, the plows pull themselves and the fried chickens fly right onto our plates." This sentence, on its own, sounds stuck amid mid-century fascinations with technology that came together at the World's Fair. However, Hilton's own foreword reveals stakes that are excluded from the *OED* reference:

> Could human beings become truly civilized if we could live in a world free of human drudgery . . . where we need no longer be afraid that the slightest slip will send us tumbling into slimy swamps, where no-one needs to swing a whip over the backs of others to keep them pulling the plow because he is afraid he might have to pull the plow himself, if he relaxes his watchfulness.[21]

While many blame cybernation for taking jobs, Hilton argues that most of those jobs are not actually worth having. Chickens may fly right onto our plates, but it is yet more significant that "in the era of cyberculture, the plows pull themselves," for automation offers release from servitude, indeed slavery, and not just a threat to employment.

Only when the book turns from substitution of machines for physical labor to discuss other areas of technological change—games, warfare, information-sharing networks, and what Hilton calls "mental work"—does it come much nearer to the objects of study in contemporary media theory.[22] Hilton subjects those functions to the same ethical consideration that she brings to bear on industrial automation. With regard to the possibility of automated warfare, she concludes:

> The computing-machine systems of the future will be much more complex than any that are in operation at present. . . . Before we permit such machines to operate in their literal and incredibly fast manner, we must be very sure that we have inserted the right assumptions and

programmed the right instructions. Above all, we must be certain that we have kept all questions that must be decided by human beings to be answered by human beings.[23]

Hilton's observations may offer a certain I-told-you-so pleasure to digital skeptics, but there is nothing shocking about the idea that human input should play an important part in human welfare. Yet two aspects of this passage might still cause surprise: one, the early date of its writing; and two, the prominent role of working-class, pacifist, and antiracist politics in the initiation (rather than in the late maturity) of the critical study of cyberculture.

Hilton's work is nearly absent from contemporary historiographies of media studies. Multiplication of institutional spaces in media studies and the philosophy of technology (both accelerating during the middle of the 1960s) means that technology and culture are now often considered together, and considered seriously. Yet while neither Hilton nor the cyberculture conference plays much of a role in those spaces, both exert direct force on the very fields that Hilton thinks worthy of human "devotion": "poetry and politics." Through principal figures of these latter fields, a radical critique of cyberculture makes its way out of the early sixties and into the experimental thinking of the next decades. In 1963, Hilton's own book appears. Then, in 1964, Hilton organizes the Conference for the Cybercultural Revolution. The conference cascades into the founding of the Institute for Cybercultural Studies at the Center for the Study of Democratic Institutions in Santa Barbara, which leads soon after to the publication of a pamphlet series called "The Age of Cyberculture," whose editorial board included Hilton and Arendt as well as Bertrand Russell and others. But the key contribution of cyberculture is less institutional than theoretical, and less pressing as an intellectual lineage than as a political demand.

The Vision in Living Form

On March 22, 1964, a letter is mailed to Lyndon Johnson from a group calling itself the Ad Hoc Committee on the Triple Revolution. This letter, a document that is sometimes called the Triple Revolution Manifesto, was signed by a luminous gang of New Left figures, including Hilton, the racial justice activists James Boggs and Bayard Rustin, the literary critics Irving Howe and Dwight MacDonald, student leaders Todd Gitlin and Tom Hayden, and assorted public intellectuals, including Robert Theobald, Gunnar Myrdal, W.H. Ferry, and Linus Pauling. The letter/manifesto demanded that Johnson recognize the connection among recent transformations in cybernation, weaponry, and human rights. For the signatories, cybernation meant full automation, weaponry meant the atom bomb and other machines of U.S. imperialism, and human rights meant both antiblack racism and civil-rights activism, which together they label "the Negro revolution." This document was a principal object of debate and dialogue at the cyberculture conference in the Terrace Ballroom of the Roosevelt Hotel, but it had already made the rounds, and later received cursory reply from the Johnson administration. It was published in the New Left magazine *Liberation*, paired with a commentary by Dave Dellinger and an interview with Paul Goodman (conducted by Roger Ebert, then undergraduate editor of the student paper). And it appeared, still later, in the pamphlet series under Hilton's supervision.

The three revolutions of the title were stitched together in the letter's final section, entitled "The Democratization of Change":

> The revolution in weaponry gives some dim promise that mankind may finally eliminate institutionalized force as the method of settling international conflict and find for it political and moral equivalents leading to a better world. The Negro revolution signals the ultimate admission of this group to the American community on equal social, political, and economic terms. The cybernation revolution proffers an existence qualitatively richer in democratic as well as material values. A social order in which men make the decisions that shape their lives becomes more possible now than ever before. . . . [However,] a vision of democratic life is made real not by technological change but by men consciously moving toward that ideal and creating institutions that will realize and nourish the vision in living form.[24]

The Triple Revolution manifesto had short-term aims for the restructuring of the work economy, as well as long-term aims for the subsequent elimination of systemic violence and inequality. In the short term, as Heather Hicks notes, the manifesto aligned with a New Left imperative to offer "the possibility that men and women might soon be able to organize their lives not in terms of a regimented schedule of formal tasks, but in response to those highly personal desires revealed by the absence of the demands of production."[25] Were this possibility to become reality, were there to be achieved what Hilton called politics and poetry, yet more radical changes would follow. Workplace technologies—computers—would likely force simultaneous transformations in the institutions and machines of war, and in the political status of African Americans.[26] These latter transformations could happen to either good or ill effect, the manifesto argued, but they would be more likely to be good (that is, to become more democratic) if automation could replace dehumanizing forms of labor with a guaranteed income and dignified noncompulsory work. Finally, and this is where it differs most from accelerationism and technological solutionism in their contemporary sense, the manifesto argued that large-scale social betterment would be a difficult long-term project.[27]

Traditions in cyberculture that flow forward from Alice Mary Hilton and the Ad Hoc Committee might be gathered under the sign of a broad-based effort at understanding and improving human relationships in the context of new technological inventions that these relationships also precipitate. The "culture" in cyberculture has a rather straightforward Geertzian sense. It is "the stories we tell ourselves about ourselves" in relation to our machines. Yet rather than merely reciting these stories, and rather than seeking less ideological stories to substitute for them, cyberculture attends to what connects those stories to one another, just as it elicits new and contrary stories. Cyberculture, in short, is a partial and associative mode of writing and reading. It is also not unsusceptible to the lure of the technological fix. Hilton and the Ad Hoc Committee undoubtedly exhibit enthusiasm or even optimism for new technology, to the extent that they link cybernation to a decrease in capitalist, racist, and militarist violence. Yet more important than a faith in any particular technology, or in any technological character to political solutions, cyberculture provides a critical diagnosis of capitalism, racism, and militarism, redefining these forces in terms of their shared components and motives.

As a result, while Hilton and the Ad Hoc Committee may well be read in terms of their prophecy, they are more profitably read through their critique, reimagining their present through a competing narrative about the past: abundance not scarcity. Their texts, and the texts that they influence, are not only futurist but also presentist. As futurist texts, they prescribe industrial and cultural practices that aim to diminish racism and end war. As futurist, they aim to heal a social body damaged by "coercion, pain, and toil";[28] to democratize technological change by urging readers to "creat[e] institutions that . . . nourish the vision in living form";[29] and to keep "all questions that must be decided by human beings to be answered by human beings."[30] These are the lessons that attach to cyberculture insofar as it names a passing moment in attitudes toward technological and political futures. Yet at the same time that these texts aim in their moment to chart an emancipatory future, they also provide transformative commentary on their own moment, the middle of the 1960s, the forking path of technological ethics.

Their responses in their own time, both figurative and presentist, point to an intersection of technological invention and human politics. They observe and refuse a common discourse in which technocratic expertise had siloed the narratives of human rights and technology, war and technology, automation and society, as if these narratives were separable. Cyberculture, in the terms of its coinage, seeks to modify the historical narrative of its own present. As a poetry and a politics, it dares anyone in any subsequent present to adopt it for use in the continued study of technology and culture, even though adopting it may mean that one must reorder one's disciplinary priorities and procedures.

Taking Advantage of a Dissensus

Dovetailing with a scholarly field that emerged during the same years, the philosophy of technology, cyberculture nevertheless flourishes partly because of that field's nascence and youthful incoherence. During the Long Seventies, a growing interest in philosophical approaches to technology coincided with a growing consensus that no existing philosophical lexicon had yet proven adequate to the topic. In short, even by the end of the seventies, philosophers of technology still had no shared idea of what the word *technology* meant. In the then-new field, there is just too much distance, too little dialogue, perhaps even incompatibility, that separates Don Ihde's influential 1979 gloss on Heidegger, wherein technology is "both the condition of the possibility of the shape of world in the contemporary sense, and the transformation of nature"[31]—or Bruce Hannay and Robert McGinn's pragmatic 1980 use of the word to name "cultural activity devoted to the production or transformation of material objects, or the creation of procedural systems";[32] or Andrew Feenberg's more semiotic definition, that same year, as "an artefact possessing a specific social meaning as well as a 'use' . . . [in the] inscription of this meaning on the natural presuppositions of social life"[33]—from Langdon Winner's definition as any "structure whose conditions of operation demand the restructuring of [its] environments."[34]

Even for such scholars as shared a discipline and published in the same venues, the greatest consistency of technological thought was that there was no consistency at all. This is a decade after the poetic declarations of "Hilton's Law," but if the politics and poetry of cyberculture survive that long, then this is why. "There is at present no highly developed philosophy of technology," Albert Borgmann explained, "because it is not clear what precisely one should disagree on. There are greatly varying views of technology, but they do not reflect a variety of explanations of a certain phenomenon: more often we have various phenomena which are called by the same name";[35] while Mario Bunge concurred that "technophilosophy is still immature and uncertain of its very object . . . which, to some, means all techniques; to others, all applied sciences (including medicine and city planning); and to still others again, something else";[36] and Winner wrote that there was no common stance "as to the nature of the problem or about the approach that an intelligent person should take in the quest for understanding."[37] This is just one field, philosophy, among many that took technology seriously in the Long Seventies. But when all that can be agreed on is that technology incites questions that are both ontological (e.g., What is that general thing called "technology"? What are some particular instances conforming to that generality?) and ethical (e.g., How should I be with this or that technology? With or without technology?), then there has not been much agreement to anything after all.

Much room remains in the public discourse, then, for claims about technology that are not philosophical in any disciplinary sense but are instead driven by imperatives of literature and activism. Cyberculture can in this way be seen to have survived better within, and at times against, the work of left critics like Herbert Marcuse, Theodore Roszak, Murray Bookchin, and Paul Goodman, all of whom wrote about technology in venues (from *New Left Notes* to the *New York Times*) that were not, or not primarily, philosophical. Typical of these thinkers, many but not all of them scholars, was an insistence that certain technologies—automation and computation above all— had only been employed for authoritarian and capitalist purposes, but that these same technologies might yet ground a theory and practice of freedom. This is the other side of a conversation in which cyberculture was engaged.

In a 1965 essay "Toward a Liberatory Technology" (written under the pseudonym Lewis Herber), Murray Bookchin determined that automation and computation, although previously employed only by centralized industry and government, would soon be appropriated by anticapitalist struggle. "With the advent of the computer we enter an entirely new dimension of industrial control systems," he wrote.[38] However, "in a future revolution the most pressing task assigned to technology will be to produce a surfeit of goods with a minimum of toil . . . to permanently open the social arena to the revolutionary people, *to keep the revolution in permanence.*"[39] Bookchin, while alert to the reactionary functions of capitalist machines, nevertheless retained hope that new technologies of automation and computation could replace an economy of scarcity with one of abundance. Marcuse, speaking in West Berlin two years later, echoed Bookchin's view of the two possible paths for technology. Marcuse argued that multiplying technologies had endangered the psychic and political lives of workers; but at the same time, "cybernetics and computers can . . . [enable] the negation of the need to earn one's living; the negation of the performance principle, of competition; the negation of the need for wasteful, ruinous productivity, which is inseparably bound up with destruction."[40]

Still, not everybody was so optimistic. In Roszak's influential book *The Making of a Counter Culture* in 1969, these technologies would likely induce more danger than benefit. Dismissing postscarcity utopia as just a leftist version of "the classic justification for technological progress . . . that it steadily frees men from the burdens of existence," Roszak predicted that "by the time we arrive at this high plateau of creative leisure, we may very well find it already thickly inhabited by an even more beneficent species of inventions which will have objectified creativity itself."[41] This, argued Roszak, would do little to fix the real problem: the continued consolidation of technical knowledge in the hands of a few specialists. For him, the technocracy, an oligarchy of experts, could be reduced through the autochthonous intellectual practices of the youth-based counterculture.

No less a youth advocate than Paul Goodman shared this rejection of the utopian view, yet did so with none of Roszak's flirtation with pessimism. Writing in 1969, in a glossy magazine for technologists called *Innovation*, Goodman admonished readers to reinvent the way technology gets talked about, and not just the way it gets used. For Goodman, after the catastrophic technologies of "Hitler's gas chambers, the first atom bombs and their subsequent development, the deterioration of the physical environment and the biosphere," there would need to be a new relation to the invention and use of machines.[42] For Goodman, the primary responsibility of researchers and designers was not to become futurists or problem solvers. Instead, with echoes of Wiener and Oppenheimer, he argued that researchers should be moral philosophers with a commitment to the general welfare. In order to produce "prudent goods for the commonweal," then, technologists would need to study "something of the social sciences, law, the fine arts, medicine, as well as relevant natural sciences."[43] This is effectively a reversal of Roszak's indictment of the technocracy. Whereas Roszak asked that the social body reduce its reliance on experts through popular regimes of technical knowledge, Goodman asked that experts put their technical knowledge in the service of the social body. The tension between these positions is as acute as the tension that would soon split the philosophy of technology into dissensus over basic terms. In both cases, partisans emerge to take one side or the other, heeding the demand to stand with or against expertise, with or against a phenomenology of technology.

The fundamental lesson of cyberculture is that cybernation might somehow enable a shift in human relations that leads finally to less exploitation.

Radicalizing this claim is George Jackson, who finds nothing to fear in increased automation, but much to abhor in capitalism. In letters from Soledad Prison written in early 1970, Jackson notes that the labor force has been "steadily decreasing and growing more skilled under the advances of automation," but that training in technical knowledge is constrained by race and class, which "casts the unskilled colonial subject into economic roles that preclude economic mobility."[44] There is no reason to defend unskilled jobs from increasing automation, Jackson writes, for these are dehumanizing and violent. Yet there is likewise no reason to defend skilled jobs, since these are unavailable to the mass of the potential labor force. Automation is fine then, at least until capitalism is dismantled. To argue otherwise is to stand by a work ethic that has only ever benefited a very few: the idea "that Amerikans [*sic*] would rather work with their hands than use a machine that could do the same work better and faster . . . sounds pretty silly to me," Jackson wrote. The technology that needs destroying, as far as Jackson is concerned, is not the automation of industrial labor but is instead the mechanisms and machinations of capitalism itself: "To destroy it will require cooperation and communion between our related parts; communion between colony and colony, nation and nation. The common bond will be the desire to humble the oppressor, the need to destroy capitalist man and his terrible, ugly machine."[45] It is Jackson, therefore, who best extends the principles of the Triple Revolution. Whereas the Ad Hoc Committee would cultivate automation as a way out of capitalism and its attendant violences, Jackson would accept automation as a necessary evil, so long as it provides time away from the assembly line in which to cultivate anticapitalist solidarity and action.

Against Scarcity

Cyberculture in the terms of its coinage starts and ends with the revolution of "poetry and politics," rather than with technologies themselves, as the aims of a technological society. In this it differs from other popular metaphors for technological change, and differs as well from more recent definitions of the word. Since the start of the twenty-first century, cyberculture has named little that is literary or even explicitly political. Around the beginning of this century, for example, media theorist Derrick de Kerckhove determined that "cyberculture is the multiplication of mass by speed . . . [whereby] even as TV and radio bring us news and information en masse from all over the world, probing technologies, such as telephone and computer networks, allow us to go instantly to any point and interact with that point";[46] and philosopher Pierre Lévy wrote that "cyberculture is the set of technologies (material and intellectual), practices, attitudes, modes of thought, and values that developed along with the growth of cyberspace."[47] Perhaps most influentially, media theorist David Silver outlined three successive stages in cyberculture, the last of which begins in the 1990s: a large-scale theoretical inquiry that Silver called "critical cyberculture studies," among whose priorities he included "the social, cultural, and economic interactions that take place online."[48] More than this, Silver writes, critical cyberculture studies will concern itself with "the stories we tell about these interactions" in both their "political and economic considerations" and also their "technological decision and design processes."[49]

These three definitions are not identical: de Kerckhove emphasizes telecommunications, while Lévy and Silver emphasize cyberspace; and Silver emphasizes the growth of a critical discipline while Lévy and de Kerckhove emphasize changing ways of life. But in contrast to Hilton's definition—cyberculture as the project of forging human community under new conditions of life and work—the three definitions are more similar than they are different. They share priorities with Hilton, but they are more about the effects of new technology and less about the material relations that emerge into and out from technology; more about the way that humans relate to technology and less about the way that humans relate to each other. In this, twenty-first century cyberculture theorists owe less to the politics and poetry of thinkers like Hilton than they owe to key phrases of midcentury cosmotechnics, like the slogans of the 1964 World's Fair, or Buckminster Fuller's one-town world, or Marshall McLuhan's global village. Such phrases imagine the transcendence of social and political difference that will follow the saturation of the world with telecommunicative and transportation technologies. By contrast, Hilton's phrase—"City of Politics and Poetry"—names an accommodation and negotiation of difference. Cyberculture, in this sense, must include a denial of the global village.

Unlike the current employments of the word, cyberculture does not require the illusion of increased proximity and communication. Instead, it takes place within and against this illusion. Following Raymond Williams, cyberculture is a practice of art and study that negotiates the "unevenly shared consciousness of persistently external events."[50] Just as it advocates technological ethics as a way to eliminate war and racism, cyberculture also opposes the doctrine of scarcity. This doctrine of scarcity obtains well into the twenty-first century, in the claim that global material resources are few, and their supplies running low, such that human rights is still often thought to be a problem of allotment and allocation, and that real patterns of consumption must change so that the human species might survive. By contrast,

Hilton and the Ad Hoc Committee insisted that world resources were more than sufficient to sustain its population, but that exploitative conditions of labor, coupled with exploitative forms of accumulation, had made resources appear scarce. Capitalism, for the latter, is suitable only for having made the machines that will enable a collective move beyond capitalism. For the former, capitalism is the shell in which individual citizens must develop more responsible practices of resource management. Whereas the doctrine of scarcity has nonrevolutionary motives that range from liberal ecology to neofascist nativism, the doctrine of abundance has emancipation as its aim.

With all this at stake, what happened to cyberculture that it never emerged as a liberatory force in the way that Hilton and the Ad Hoc Committee, anarchists like Goodman and Bookchin, and ultimately George Jackson might have imagined? Could cybernation ever have curtailed the growth of dehumanizing low-skill labor, beaten the myth of scarcity, and given way to emancipation, to racial justice and pacifism and the end of capitalism, in the way that its advocates had hoped? In *The Hidden Injuries of Class* of 1972, sociologist Richard Sennett says no. Even in the "purely 'technological' industries like the computer field," Sennett laments, "the efficiency in production brought by advances in technology has not eliminated low-level routine jobs but has rather shifted them around and, in aggregate, perhaps, even increased them."[51] Speaking retrospectively that same year to the National Science Teachers Association, Hilton appears to accept this disappointing fate for the cybercultural project, albeit with a different diagnosis of cause: "Some of our prognoses have since been verified by events. Most of them have not because one of our assumptions—massive investments in nondefense industries—has not been put into practice."[52] Hilton insists, beyond Sennett, that cybernation fails to fix everything not only because its owners want it to fail, through a kind of industrial inertia, but also because automation remains triangulated with the other two technological revolutions: war and human rights. It follows for her that if one of those revolutions fails, and the modification of military technology has certainly failed, then the other two revolutions will fail too.

As if in a last-ditch effort, Hilton insists of her moment: "There was never before a time when man had the tools to create true abundance on earth and leisure for all human beings and the ability to educate all so that we can take our chances on the miracles mankind might wreak when all human beings could develop all the potential wisdom and goodness that may be in them."[53] With a final sigh, Hilton admits that in spite of the advantages of the moment, wisdom and goodness are not likely forthcoming. Perhaps some cybernation has contributed to some reduction in poverty. Perhaps some amelioration of poverty has contributed to some amelioration of human rights. But the war in Vietnam still has not ended. The global distribution of resources remains in a condition of enforced scarcity. The technological character of racism is disavowed. War remains a means to accumulate wealth. Little that is structural has changed. Hilton concludes: "No human being can be good and kind and law-abiding with the wolf at his starving children's cave."[54] There is a striking tension between this sad acceptance of reality and the persistent hope she still holds out, for "miracles mankind might wreak." The mode of transition from one to the other, from starvation to a world without wolves, is a speculative distortion that turns reality into miracle.

CHAPTER 4

Distortion

Through our scientific and technological genius, we have made of this world a neighborhood and yet we have not had the ethical commitment to make of it a brotherhood.

—Martin Luther King, Jr., at the National Cathedral (1968)

In its coinage, *cyberculture* named the potential for human poetry and politics, even amid dehumanizing machines. But what is cyberculture that it could enable any such thing? The critics of cyberculture wrote what appear now as prescient and potentially paradigm-breaking texts—opposing war and racism and capitalism at once, with a complex understanding that they are connected—yet the texts themselves have been largely neglected. As futurist texts, indeed very few lessons may be drawn from them. As instances of futurism, the historical genre of midcentury prose, they are merely interesting and sometimes exemplary. But as commentary on their present, on the embattled ground of technology and culture in the decade and a half after the midsixties, they teach much more. In this chapter, I show how the idea of cyberculture was absorbed into practices of activist and literary writing, expanded and extemporized by seemingly dissimilar theories and practices of dismantling. Ranging from poetry and prose fiction to sermons and feminist polemic—from Thomas Merton to Shulamith Firestone to Philip José Farmer to Martin Luther King, Jr., to radical literary theorists—the inheritance of cyberculture goes beyond futurist hope and prediction. Cyberculture, in its survival, is literary: a speculative attitude, an urgent disfiguration of the way things are, a gathering insistence on the material interconnections among racialized, gendered, and technological violence. Whether they are poetry or prose, fiction or nonfiction, the writings of cyberculture thus align with a regime of reading that Samuel R. Delany calls the significant distortion of the present.

In discussions of science fiction during the same years, Delany finds that a narrative of not-yet-existing technology is "neither one of prediction nor of prophecy" but instead "one of dialogic, contestatory, agonistic creativity . . . in a significant distortion of the present."[1] Distortion, too, is dismantling. It is a blurring or erasure of the continuous narrative of technological progress by literary and speculative thought. As a significant distortion of the present, the claims for a cybernated future are not "about the future" at all, but instead exert all their pressure on incipient and ongoing changes to human life in the final third of the century. Cybercultural texts require their readers to believe what every available datum would call impossible: that planetary resources are abundant rather than scarce; that automation is a potential benefit, and not only a plague, to workers; that racist and militarist violence might be alleviated through wholesale structural change. As a source of "agonistic creativity," cyberculture "contests" the commonest presumptions about technological emergence: the technophilic presumption that machines enable greater efficiency and therefore greater happiness, as well as the technophobic presumption that machines replace human labor and thereby diminish human dignity.

Ursula K. Le Guin defines science-fictional world making in a way that, echoing Delany, grounds generic practices in the transformation of an existing world. Rather than imagining the capacity of a genre to invent new worlds elsewhere, in the future or in outer space, Le Guin insists on a continued focus on the writer's own existing world. In a 1980 interview, she asks: "What about making the world new, making the world different? That is the mark of the political imagination. Then you've got utopia, dystopia or whatever. But what about making the not-new? What about making this world, this old world that we live in? . . . In a day-to-day living sense, we make the world we inhabit."[2] Were cyberculture to be considered as a variety of futurism, it might provide hope that new forms of technology will eventually liberate humanity from its labors, or it might be cited as a literary genre in an intellectual history of liberatory thinking about technologies to come. But when read as a significant distortion of its present, or as a way of making "the world we inhabit," in a "day-to-day living sense," cyberculture provides something else entirely: a theory.

Such a theory aligns with Alexis Shotwell's repurposing of speculative thought for radical philosophy: "There is not a single pure or perfect future toward which we stretch. . . . The shimmer here between the necessity of imagined tomorrows and control of the too-quickly-arriving tomorrow is the space of the kind of creativity signaled by prefigurative political practice."[3] A proper theory of cyberculture, one imbued with Shotwell's prefigurative creativity, must argue that technology has never simply been applied to life, with secondary implications that are racial, classed, or militarizing. Instead it must argue that race, class, and militarism are formative conditions in the existing world that foster technology and persist in it. Shotwell's too-quickly-arriving tomorrow is a synonym for those technologies of centralized power (e.g., guns, prisons, and industrial robots) that are all built from the same parts, and from the same political conditions, such that the fates of militarism, racism, and cybernation have only ever been knotted together. Cyberculture thus calls for a cultivation of imaginative practices, whether literary or activist, for cutting through the knot.

The Large World House

Speaking in the National Cathedral in Washington in late March of 1968, Martin Luther King, Jr., spoke about this knot. In this, his final sermon, four days before his assassination, King implored his listeners to "remain awake through a great revolution." The phrase was the title of the sermon, and makes reference to Washington Irving's character Rip Van Winkle, who slept through the most significant alterations of his world. The rest of the sermon concerns the discourse of cyberculture directly. For King, the Americans of the 1960s have been "peacefully snoring up in the mountain [while] a revolution was taking place that at points would change the course of history." He continues:

> In a sense it is a triple revolution: that is, a technological revolution, with the impact of automation and cybernation; then there is a revolution of weaponry, with the emergence of atomic and nuclear weapons of warfare; then there is a human rights revolution, with the freedom explosion that is taking place all over the world.[4]

King here paraphrases the manifesto of the Ad Hoc Committee, which he had likely learned about from Bayard Rustin, a signatory to it. For King, in his reading of the document, new technologies demand an ethics of commitment and community rather than a technics of automation and communication.

King does not deny the importance of technical matters, and neither does he refuse the technologies themselves. He seems to welcome them, in fact, while refusing to see any inherent contribution by these technologies to the conduct of human community. Any technological development, for King, should be supplemented by a culture of mutual care:

> The geographical oneness of this age has come into being to a large extent through modern man's scientific ingenuity. Modern man through his scientific genius has been able to dwarf distance and place time in chains. And our jet planes have compressed into minutes distances that once took weeks and even months. All of this tells us that our world is a neighborhood. Through our scientific and technological genius, we have made of this world a neighborhood and yet we have not had the ethical commitment to make of it a brotherhood.[5]

The much-heralded connectivity of the world is ostensibly brought into being by the advances in telephony and transportation ("our jet planes have compressed [distances] into minutes"), as well as by radio, television, and other electronic media. And yet this neighborhood, for King, has cropped up without any consideration of consequences, which are still up for grabs. The neighborhood is an event only, a neutral happening, something that has occurred, obliging neighbors to "remain awake" to its influence, but without engendering any of the celebrated social effects that are often associated with proximity. King's reaction to this event is neither to cheer nor to deride the new closeness, but instead to demand a complementary gesture: brotherhood, a new ethics of shared feeling and obligation.

The shape and scale of this obligation are worked out, and techniques of planetary brotherhood are brought into contact with thinking machines, in King's final book, *Where Do We Go from Here?* The book is published in 1967, and its last chapter, entitled "The World House," is essentially a longer, looser, early version of the 1968 sermon. Of jet planes, medical science, and cybernation, King writes: "All this is a dazzling picture of the furniture, the workshop, the spacious rooms, the new decorations and the architectural pattern of the large world house in which we are living."[6] The large world house predates its new furniture—and it does seem to be very nice, very useful furniture—and the large world house will survive subsequent redecorations. If this furniture is synonymous with the technologies of freedom and warfare, as mentioned in the sermon, then the next question is neither *How have we been changed by our furniture?*, nor *What furniture should we buy next?* Instead, King's title asks *Where Do We Go from Here?* It asks how will "we" live in the large world house that we inherit, furnished?

> These are revolutionary times. All over the globe men are revolting against old systems of exploitation and oppression, and out of the

wombs of a frail world new systems of justice and equality are being born. . . . Our only hope today lies in our ability to recapture the revolutionary spirit and go out into a sometimes hostile world declaring eternal opposition to poverty, racism, and militarism.[7]

King here returns to the three principal terms of the Triple Revolution: poverty, to be alleviated by cybernation; and then racism and militarism, to be alleviated by relief from poverty and a rethinking of weaponry.

Orderly Recomplication

As King redraws the architectural plans of the political world—the amplified ideal of the world house—he imposes something like Samuel Delany's speculative formula, the significant distortion of the present. In 1978's *The American Shore*, Delany argues:

> Science fiction is not *about* the future. It uses the future as a convention to present a significant distortion of the present. But the form of the distortion is covered neither by the exaggerative transformations of satire, parody, or lampoon; nor by the reductive transformations of fable, fairy tale, or fantasy: its essential transformative methods are random combination and orderly recomplication.[8]

For Delany, science fiction is only secondarily a genre of cultural production, and far more importantly a mode of reading. And as a mode of reading, it is characterized by "random combination and orderly recomplication." It is this sort of reading that King performs. With the cosmotechnic metaphors of media theory entering wide circulation, King insists on a different spatial arrangement that would subordinate neighborly proximity and enhance communal love.

The world house is the orderly recomplication of the global village, a narrative of change without "fairy tale or fantasy," and a distortion of the present. What cyberculture shares with media theory and cybernetics is, to quote N. Katherine Hayles, the latter's "disturbing and potentially revolutionary . . . idea that the boundaries of the human subject are constructed rather than given."[9] Yet whereas media theory achieves this indistinction by assuming a basic similarity among human beings brought into proximity, and whereas cybernetics achieves it by questioning the status of observers in communicative systems, cyberculture assumes untraversable forms of social difference while remaining committed to ethical responsibility and indifferent to matters of empiricism. No observer is necessary, indeed none is warranted, where the investment in communicative systems is largely figurative and generic. King's argument hinges not on the technologies of communication and travel that should be celebrated, but instead on the technologies of "poverty, racism, and militarism" that might yet be reinvented, transformed, or destroyed. Media theory and cybernetics see thinking machines as having tested boundaries of the human because their focus is on real technological invention. In King, thinking machines are an excuse to talk about something else. Human boundaries are tested through the force of a metaphor—the world house in which humans and human imperatives are movable furniture—and the demand for care in a world of machines, irrespective of how those machines work.

When Delany describes his way of reading, he departs from the sanctioned and disciplinary modes of reading that are associated with literary study. In the context of black literature and racial justice, Madhu Dubey has noted that Delany pairs significant distortion with a second related proposition: technology allegorizes rhetoricity. Dubey explains: "Science fiction forces us to take technological metaphors for social reality seriously, often by literalizing them, and thereby paradoxically clarifying their workings *as* rhetorical devices. When technology allegorizes rhetoricity, readers are alerted to the fuzzy political work often performed by technological metaphors."[10] So it is in King's speech, which is not about the conditions of a technologized future but is instead about the conditions of emancipation for which technology serves as a poetic or rhetorical figure. Whatever their prior investment in technology, readers and listeners must take technological metaphors seriously if they are to appreciate the moral and ethical value of this world house. Irrespective of whether dignified craft may ever replace exploitative labor, or whether abundance may ever replace scarcity, the metaphors hold their force. This is how, in Dubey's terms, the metaphors become legible as metaphors. King's technological metaphors enable readers and listeners to engage in the "fuzzy political work" of emancipation, because the readers and listeners know that this work is, in part, the work of words.

The distorted present has no content but its transformed inheritance from the past. King thus pursues an active rearrangement of the narration of the past, such that the past may furnish a present community with collective responsibility for what comes after. He imagines the world as a house to be lived in, agonistically and contentiously, rather than a distance to be traversed by commodities or messages. This rearranged present demands another way of reading, as Delany argues, against the sanctioned modes: not the analysis of "subjective time laid down through real history [as on] the road," but instead the capacity of science fiction to "open up that highway into a boundariless plane . . . which [in turn] quickly deliquesces into a roiling ocean of possibilities."[11] Politics and poetry emerge in a roiling present, legible to a nonlinear form of analysis, driven by urgent demands for humans to care for each other, and for the large world house, for the duration of their residence.

Back Then . . .

In 2013, Melinda Gates addressed graduating students at Duke University's commencement ceremony. Her point of departure is "Remaining Awake through a Great Revolution," the final sermon of Martin Luther King, Jr. The address, itself a sort of sermon, is an extended riff on King's words: "Through our scientific and technological genius, we have made of this world a neighborhood and yet we have not had the ethical commitment to make of it a brotherhood." To this line, Gates replies:

> With 50 years of hindsight, I think it's fair to say Dr. King was premature in calling the world a neighborhood. Back then, Americans lumped whole continents into something they referred to as the Third World, as if the people on the other side of the planet were an undifferentiated mass whose defining feature was that they were not like us. But as a result of the ongoing communications revolution, your world really can be a neighborhood. So the ethical commitment Dr. King spoke of is yours to live up to.[12]

Gates's misunderstanding is instructive. She betrays no knowledge of King's literary and philosophical intertexts, but instead steers King energetically, transforming him into a prophet of a Silicon Valley techno-futurism that he could not possibly have predicted. The purpose of her commandeering is to declare provisional victory for the present moment. Technical connections having been made—the "communications revolution" having effectively shrunk the world—it only remains to execute those connections according to principles of mutual responsibility. Wearing the ethical face of cosmotechnics, Gates refuses to allow the "mass" of the world population to remain "undifferentiated" by global capitalism. Her vision is fundamentally inclusive. As such, it excludes what it cannot understand: laboring conditions that are marginal to conditions of capitalist production and exchange, as much in the ostensible First World as in the ostensible "Third."

When Gates points "back then" to those who "lumped whole continents into . . . the Third World," she encourages a triumphalism of a technological present from which her family prodigiously profits on this side of the world. When she notes the "ongoing communications revolution," she repurposes for marketing reasons a noun, *revolution*, that, in King's texts and intertexts, denotes a genuine social overturning. Moreover, what she markets with this noun are her own products, especially the Gates Foundation's brand of philanthrocapitalism, that repeat the very forms of colonialism and exploitation that she has disavowed by consigning them to that supposedly bygone era. Gates's most damaging misreading may belong less to her material interests in Microsoft and the Gates Foundation and more to her moment, the early twenty-first century. She says, "Dr. King was premature in calling the world a neighborhood," but that, of course, is exactly what he did not call the world. The world, for King, was a large and capacious house, furnished and patterned by technical means, but ever under construction and redesign, and in want of human care. In the line that Gates quotes, King does not celebrate either "technological genius" or "neighborhood" (as she presumes, with her "50 years of hindsight," that it does). Rather, King makes ironic use of a common metaphor, neighborhood, best set aside for those who would celebrate mere proximity. In its place, King provided another metaphor, brotherhood, that may better sustain an ethical imperative.

Amid claims to the reality of the global village, and against the backdrop of the dedication of the 1964 World's Fair in Queens—"Man's Achievement on a Shrinking Globe in an Expanding Universe"—King's call for brotherhood is an insistence that mere proximity, even if it were possible, would not be enough. Brotherhood, by contrast, would be enough. But would it, really? Few voices have been heard so widely and for so long as King's, yet still no universal "ethical commitment" has arrived to end the violent exploitations and exclusions of our species. King prescribes, and does not describe. He projects a postideological social vision that at least partly repeats the ideological formations of its moment. It hangs on a gendered term, "brotherhood," that can at best ever unite one half of the species; and it can only imagine a continued centrality of automation and telecommunication in the conduct of human life. This is not atypical of perspectives in cyberculture during the final third of the twentieth century. Like the rest of that moment, which is never quite a movement, King therefore calls for a double response: one, its rereading in the context of cybercultural poetry and politics, especially what has been lost or ignored; and two, along with those texts, its transcendence or distortion in a present that is, after all, so very different than it was.

Dreaming into Action

Just as King's reading of cyberculture comes to resemble science fiction, so cyberculture also influences and includes science fiction. The most explicit debts to cyberculture, in that genre, belong to Philip José Farmer. The

novelist of the *Riverworld* and *World of Tiers* series makes numerous efforts
to introduce the Triple Revolution letter to the genre's readers and writers.
The first of these efforts is in "Riders of the Purple Wage," a Joycean pun-
filled novella about the inevitability of unregulated perversion in a population with a universal guaranteed income and lots of free time. With the
novella's first publication, in the 1967 anthology *Dangerous Visions*, editor
Harlan Ellison attaches an afterword by Farmer, emphasizing the influence
of the Triple Revolution letter:

> The writers of the document know that mankind is on the threshold
> of an age which demands a fundamental re-examination of existing
> values and institutions. . . . This document may be a dating point for
> historians, a convenient pinpointing to indicate when the new era of
> "planned societies" began. It may take a place alongside such impor-
> tant documents as the Magna Carta, Declaration of Independence,
> Communist Manifesto, etc.[13]

Clearly there is no canonical status in store for the Triple Revolution letter—
no list anywhere will place it among these other "important documents."
But the letter had its effect on Farmer.

Giving the keynote speech at that year's Hugo Awards, after having picked
up his own Hugo for "Riders of the Purple Wage," Farmer devotes his entire
speech to explaining and promoting the Triple Revolution. Chief among
his priorities were the unregulated and as yet unpredictable forms of sexual attachment that might be explored in an economy of abundance rather
than scarcity. For Farmer, this theory is necessary to any new discourse of
media theory, because of the limited usefulness of that field's major figure,
Marshall McLuhan. For Farmer, McLuhan's texts lack a method, and any
correctness there (in claims about the global village, but also those concerning the medium as a message, and art as a history of the future) is more or
less accidental: "To prove his theses, quite often he strains the bowels of
his mind; you can hear the grunts and groans, and the result is flatulence.
Despite which he must be listened to. He is three fourths right."[14] With this
scatological caveat, Farmer sets aside those more influential perspectives on
the new technologies of labor and communication and shifts toward Paul
Goodman and to the Triple Revolution letter. Farmer asks, "Do you want
our children, and our grandchildren, to inherit a stinking, suffering, perhaps
doomed, world? Doomed to choke in its own waste products—mental, emotional, and certainly physical?"[15] How will the future inherit a wrecked present? Farmer's keynote is not a reversal of one ideological tenet, as is cyberculture's reversal of scarcity into abundance, followed by a detailed program of

action. Instead, Farmer makes a wholesale distortion of the present, which he sees figuratively as a doomed pile of "waste products," and then poses a question about it. With his feet in this pile of present, Farmer bids his audience of science-fictionists to join in a collective redrawing of a fully articulated alternative vision: "You—the science fiction people—have always dreamed of the future. . . . Now you are the 'Fertile Void' . . . ready to convert dreaming into action. And you have a long-standing—if loose knit—effectively operating group."[16] No such collective reimagining seems to have happened. Farmer proposed alliances that were never assembled, and publications that were never printed. And yet Farmer's own prose is increasingly guided by the texts of cyberculture.

Farmer's best-known series of novels, the *Riverworld* books, begins from the question: If a new technology were to replace scarcity with abundance, then what style of human life, what kinds of human love and war, would follow? This series of novels, beginning with *To Your Scattered Bodies Go* in 1971, imagines that every human, living or dead, has awoken on a distant planet. The custodians of this other planet, the watery world of the series title, have provided humanity with many copies of one device (called a "grailstone") that can dispense ample food to whole communities of transplanted inhabitants, and many more of another device (called a "grail") into which the food is dispensed. Neither device need be fully understood before it begins to license Farmer's fictional exploration of the love, art, and warfare that humans might engage in, where scarcity is no longer at issue. Leslie Fiedler, a contemporary of Farmer, notes that the food-producing grailstone is no more than "a kind of portable short-order kitchen provided by the invisible masters of a warmed-over universe," and the generic devices of science fiction themselves only provide Farmer with "a warrant for constructing Universes of his own."[17] The point of Farmer's series is not to draw up technical specs for the machine that ushers in an era of abundance. Science fiction is only a "warrant" for Farmer—a way of finding an audience for principles of the Triple Revolution. More importantly, it is a way to distort the present and, through such distortion, foreground the potential of thinking through cyberculture. This distortion refuses all of the technological solutions that are already available. As Delany argues at the end of the decade: "When our new computer age is splitting social classes thought to be homogeneous into technophiles and technophobes . . . SF, rather than expressing this growing split, offers the most persistent integrative message."[18] It is this "integrative message," neither technophilic nor technophobic, that is practiced by *To Your Scattered Bodies Go . . .* when it tells of the conflict and care between heroes and villains of the human past.

The central conflict of the book is between Sir Richard Burton and Hermann Göring, and its central alliance is between Burton and Alice Hargreaves, the historical model for Lewis Carroll's *Alice in Wonderland*. The book thus stages dialogue, contestation, and agonism not only between characters but also between scattered literary traditions and between cultural and political forces: masculinism and feminism, British and Nazi imperialisms. In excess of these dynamics, the book focuses on the grail and grailstone, the machines that permit life to continue on a planet with few other resources. As in Delany's diagnosis, Farmer's novel refuses the separation of community into technophilic and technophobic classes. Instead, Farmer offers an "integrative" paradox: in order to understand the life-preserving machine, one must make use of it, or else perish. There is no optimistic future in this vision, but only a conceptual knot to be cut through, a present to be distorted. As Farmer writes, it is not utopian: "Not unless you literally translate Utopia, which means Nowhere. Nowhere is a station on the road to Somewhere, and Somewhere is never as good as it was cracked up to be. Despite which I say, Bon Voyage, Triple Revolutions!"[19]

Total Redefinition

Among the many writers who absorbed the critical language of cyberculture in the Long Seventies, and who touted radical uses for cybernation, one of the most influential may be the feminist theorist Shulamith Firestone. For Firestone, cybernation was the keystone of a wholesale change toward a revolutionary antifamilialist ecofeminism. For her, moreover, there is no hope for a universal income, or for an increase in leisure time, with the arrival of total automation. Instead, modeling a way forward with a richer critical apparatus, and with clearer relevance to political movements in the twenty-first century, Firestone argues that an increase in workplace automation will lead to unemployment and anger, until women grab hold of those very cybernated means of production:

> A feminist revolution could be the decisive factor in establishing a new ecological balance: attention drawn to the population explosion, a shifting of emphasis from reproduction to contraception, and demands for the full development of artificial reproduction would provide an alternative to the oppressions of the biological family; cybernation, by changing man's relationship to work and wages, by transforming "work" to "play" (activity done for its own sake), would allow for a total redefinition of the economy.[20]

Firestone's tactics, the seizure of the means of production and the elimi-nation of the gendered division of labor, are more radical than the tactics of other cyberculture writers. Yet Firestone, like those writers, does speak of revolution. Moreover, her vision (a postfamilial domestic economy, a national and global economy that permits play and ecological balance) does not contradict theirs, but only shifts in emphasis.

Unlike them, however, Firestone has received ample scholarly response by literary and media critics. This response in turn illuminates not only Firestone but also the rest of cyberculture. Susanna Paasonen, for example, acknowl-edges the difficulty of retrieving Firestone for contemporary digital femi-nism. Paasonen argues that theoretical models can be seen to lose traction over time: "As texts are situated in a reductive opposition toward each other, the more recent ones can be posed as the 'new and much improved' version of feminist theory . . . [and] earlier work easily appears quaint."[21] Moreover, for Paasonen, Firestone's "blunt, seldom ironic, incessantly passionate" voice is nearly silenced, but no less necessary, in a moment when other kinds of feminist voice speak more loudly.[22] Something similar might be said about cyberculture: in a time when cyberculture denotes not what Hilton desig-nated in her coinage, but instead the rapidly multiplying, swiftly outmoded platforms and experiences of digital content, it can sound quaint to talk about automation. Still, quaintness is not the only risk of cyberculture. As Nina Power has written of Firestone, the cybercultural vision is undermined as much by its drift toward technological determinism as by its "free-floating utopianism."[23] The same can be said of Hilton and the manifesto—that while they might not accept automation or telecommunication as a necessary good, they would nevertheless insist that these technologies could yet be used for good. In spite of Firestone's avowal of a determined techno-future, Power argues, her critique can yet serve in the present "as a useful corrective to the idealist excesses of contemporary theories of cyberspace and immaterial labor."[24] With Power's reply to Firestone in mind, a slew of new questions can emerge for anyone asking why cyberculture, or its scholarly apparatus, should invoke one set of priorities and not another. Why, for example, should computation now be defined by the reach of social media rather than by an ethical celebration of mutual responsibility and care? Following Power's con-clusion, but indeed following the whole tradition of which Firestone is only a late exponent, if the "technological revolution is to be preserved, it should be in this most joyful of modes."[25]

To develop this joyful mode without disavowing its quaint optimism, the answer is again to treat cyberculture not only as futurist, but also as a distortion of its present. To read Firestone's text as a distortion of its present is to note merely that if other technologies are possible, then other social practices, even practices that are rarely considered in relation to new technology, may be possible too. In fact, to consider *The Dialectic of Sex* in connection to science fiction is to renew the spirit of the book's early reception. Writing in 1971, in the magazine *Fantasy and Science Fiction*, science-fiction feminist Joanna Russ gave *The Dialectic of Sex* a thrillingly positive review. Insisting that "you will have a hard time with this book if you believe that Capitalism is God's Way or that Manly Competition is the Law of the Universe—but then you can go back to reading the *Skylark of Valeron* or whatever," she calls Firestone's book "the most exciting social extrapolation around nowadays."[26] Moreover, Russ concludes, Firestone offers "fascinating alternatives to the family which recall much of Samuel Delany's fiction," and that are actually practicable "within a century, she says, if we don't blow ourselves up."[27]

Characterizing *The Dialectic of Sex* as science fiction, for Russ, means including Firestone in the late phase of the genre's so-called New Wave (of which she, Russ, was a prominent practitioner). In a 1971 discussion of the New Wave, Russ argues: "The only thing that makes many stories science-fiction is that they are not about things as they are."[28] Any story can be science fiction, she writes, so long as it concerns "things as they may be or might have been."[29] It is vital to negate the known present on behalf of a possible present. To tell of things that *may be* rather than things that *are*: this is to depart from a political moment without imagining its alternatives in positive terms. Russ's criteria are fully consonant with Delany's "significant distortion of the present." Whereas Delany would ignore the future as the necessary outcome of a known present, so that he might explore the supersaturation of the present by the past, Russ would reverse the procedure to the same end: negating the present as it is known, so as to stage a refiguration of a displaced past. Both refuse any future that might be imagined from the compromised tools of our own time, and instead seek to build an active and ongoing signifying practice that is guided by restlessness, the rejection of inherited categories, and the condensed air of moments of something else. Keeping in mind Delany and Russ, Firestone and her critics, it gets easier to get hold of this something, and to grapple with the vexed emotions that characterize all of cyberculture: quaint yet unquestionably relevant; naively futurist yet transformatively presentist; optimistic yet also soberly dedicated to the hard work of making the world into something that is not what it is.

Literary Politics

Not only is the political critique of technology shot through with literary thinking during these years. Political thought about literature is equally embedded with a critique of technology, coming to a head at the 1968 convention of the Modern Language Association, in the Americana Hotel in New York City. There, five pasteboard posters had been hung to protest the MLA's explicitly apolitical aims, by scholars attempting to bring principles of emancipation into the teaching and writing of literature. It was Florence Howe, an assistant professor who would later cofound the Feminist Press and serve as MLA president, who hung the posters. "Let's put humanity back in the humanities," proclaimed one poster; while another, quoting William Blake's *The Marriage of Heaven and Hell*, read: "The tigers of wrath are wiser than the horses of instruction." When hotel staff attempted to tear down the placards, three conferees stood in the way. Police arrived and arrested the three conferees, including Louis Kampf, an associate professor who was himself a future MLA president. According to the *New York Times*, Kampf and the others were charged and held for "the defacement of the hotel's white pillars by the transparent adhesive tape that fixed the messages there."[30] In the wake of the arrest, the tenor of the conference grew heated. The association's business meeting went long past its allotted time, as this radical caucus brought resolutions that echo the concerns of the Triple Revolution manifesto, protesting "about the war in Vietnam, about racism in America, and about the plan to hold next year's conference in Chicago"—a city whose police had become infamous for their violent response to leftist protest.[31] Alliances were forged in that meeting that led to longstanding initiatives in the areas of women's studies, black studies, and Marxist theory.[32]

By citing this event in a discussion of speculative cyberculture, I do not mean to say that it is itself science fictional, or that the culture of technology is its main concern. Yet in the acts and works of the MLA's radical caucus, there is skepticism of technology and technocratic institutions, and a combat against racism and militarism, that are perfectly consonant with the skepticisms of cyberculture. Moreover, in targeting the mythos of literariness, as they did in lectures and articles after the 1968 convention, members of the caucus do recall Delany's criticism of the academically sanctioned (i.e., non–science fictional) genres of fiction. Where a scholarly discipline undercuts its central object without bringing itself to a close, it has followed Delany in embarking across "a boundariless plane . . . which [in turn] quickly deliquesces into a roiling ocean of possibilities."[33]

At the next MLA conference, in 1969, Frederick Crews described the role of advanced literary study in helping to legitimate the changing shape of empire: "When capitalism moves . . . from acquiring basic industries within a home nation to imposing an international order in which the desired flow of raw materials and sales is encouraged by armed force, an advanced school of thought arises favoring state regulation, foreign aid, placation of the unemployed, and similar measures to help things go smoothly."[34] To be sure, Crews did not credit literature or literary studies with the entirety of this legitimation. Nevertheless, the point was made: by not explicitly refuting the conditions of social and political exploitation, liberals in literary studies had lent comfort to their ideological enemies. A similarly hard line was taken by Bruce Franklin at the same MLA conference in 1969. Franklin argued that literary training is a training in bourgeois values: "The historic mission of the scholar-critic-professor of literature is none other than the shaping of values," such that "at this point in the crisis of bourgeois ideology" a professor must teach "that the *study* of great literary achievements is more significant than taking social or political action."[35]

Along similar lines, yet with his eyes set on increased scholarly enthusiasm for new machines, Alan C. Purves told the assembly at the National Council of Teachers of English that same year: "Teachers in the humanities have become technocrats or technophiles, and thus are less concerned with the problems we all face than are the social scientists."[36] To these members and allies of the MLA's radical caucus there could be nothing inherently revolutionary about poetry and fiction. Literature is not radical by virtue of being literature, and literary instruction runs the risk of serving the very "technocrats and technophiles" who, to Purves, had themselves only ever served capitalism and empire. Poetry and fiction, on this view, provide ideological cover for mechanisms of industry and state that are racist, sexist, technocentric, and militarist. The radical caucus thus embodies a sort of outer edge to the discourse of literary politics, where literary thinking is perceived to be evocative and illuminating but not exceptional. From that outer edge, literary discourse must turn to oppose the technocrats and technophiles at any cost, even if it means opposing literature and literary study. To deny the collegiality of "technocrats and technophiles," as does Purves; to refuse to smooth out the alignment of armed force with the technologies of international trade, as does Crews; to contest complicity with militarism and racist policing, as was done in the 1968 business meeting: these acts become intelligible within a literary politics of cyberculture, even as they endanger dearly held literary ideals. Without a defense of literature as such, the radical caucus nevertheless plants its flag within literary studies, as a least worst site for solidarity against technological forms of exploitation.

By later in the decade, Joanna Russ would double down on this radical skepticism. Whereas the radical caucus had put literary instruction in the crosshairs of a critique, Russ extends the critique into a polemic against any practitioner who neglects to wield literature against capitalism and American empire. At still another convention of the MLA, Russ pushed to decenter technology as a primary object of literary and cultural analysis: "Technology is a *non-subject*," she argued: "the sexy rock star of the academic humanities, and like the rock star, is a consolation for and an obfuscation of, something else. Talk about technology is an *addiction*."[37] For Russ, technology had become all that anyone (at least in her cohort of English professors and science fiction writers) ever wanted to talk about. It would be folly, she argued, to ignore the concrete effects of technological change on artistic and political ways of responding to the world. But it was an equal error to make the technologies of communication or computation into the principal objects of a creative or scholarly myopia. Russ concluded: "The technology-obsessed must give up talking about technology when it is economics and politics which are at issue."[38] Russ demanded, in short, that literature and theory may concern themselves with technology, but only insofar as technology is a term of economics and politics (rather than the other way around). For Russ, technology must be decentered in scholarship and culture, which should learn to demythologize technology and refuse to accept technology as a stand-in for political and economic concerns. For her, the language for decentering and demythologizing technology is to be found in literary scholarship, activist argument, and science fiction. This is how, it can be remembered, Russ says we can negate the known present on behalf of a possible present. Moreover, political writing about racism, heteromasculinism, and militarism can be read in similarly speculative dimensions, in pursuit of things not as they are.

The prose and poetry of social justice departs from its political moment without imagining its alternatives in positive terms, with an orientation that is continuous with science fiction. In 1966, it was Bruce Franklin who had written that science fiction was a technique for "living in the era of the Triple Revolution,"[39] and so it would become in the decade and a half that followed, to the degree that "one good working definition of science fiction may be the literature which, growing with science and technology, evaluates it and relates it meaningfully to the rest of human existence."[40] Yet at the same time, epistemologically speaking, the Long Seventies bring challenges to that dividing line between human and technology. Teresa de Lauretis defines the genre tellingly on the other end of that period, in 1980: "The historical and generic specificity of SF . . . is to be sought in SF's re-literalization of language, its way of posing the relation between human figure and physical / material / technological ground, its ability to . . . produce other conceptual models of human reality."[41] What activist literary theory shares with science fiction, in short, is a way of expressing what is not but might be or should be, refiguring and reconfiguring the field of the possible. In this way, as Donna Haraway famously argues in 1985, "the boundary between science fiction and social reality is an optical illusion."[42] Reading cyberculture through these lenses, as a science-fictional distortion, it might then be said to name the spread of machines as these become more meaningful to human relations, yet simultaneously to name the newly problematic human / machine distinction.

Edifying Cables

As distortion of the present, and as fuel for literary activism, cyberculture can be seen to spread out across literary production after the middle of the 1960s. As a poetic practice, it is likely that much flows forward from a single moment in 1964, when James Laughlin devotes an entire introduction of *New Directions in Prose & Poetry* to the Triple Revolution. Writing of himself in the third person, Laughlin notes that this is the first editorial note that he has written in many years. While early issues are prefaced "with personal and ingenuous exhortations to the improvement of the social order," Laughlin acknowledges, he later "began to feel foolish about preaching to people who doubtless had more sense than he did."[43] There had been no introduction in a long time, and it would be a long time before another would appear. But in 1964, Laughlin cannot help but "preach" that a tremendous prosperity for some, in the United States, has not spared others from poverty. Adding a fourth revolution—population growth—to the list devised by the Ad Hoc Committee, Laughlin wonders:

> A continuing inflation in our own economy, which has brought prosperity to the majority of US citizens but still left substantial minorities in extreme poverty . . . , has distracted almost all of us from questioning the workings of the American economic machine. . . . Four interrelating factors: 1) Population growth; 2) Automation; 3) The Negro Revolution; 4) The Arms Race make imperative a fundamental reassessment of the economic system if we are to avoid crises and perhaps catastrophe.[44]

None of the contributions to the issue of *New Directions* was invested in this intersection, which exerted no force over the editing of the issue. Laughlin's intervention goes nowhere, and ends precipitously, before it can move from the borrowed diagnosis to anything like a political or poetic program.

However, one of the contributors to that volume of *New Directions* is the monk and poet Thomas Merton, who does provide such a program. Merton and Laughlin appear to have learned about the Triple Revolution from the same member of the Ad Hoc Committee, W.H. Ferry. In a letter to Ferry just one day after the manifesto was mailed to the White House, Merton writes: "'Triple Revolution' is urgent and clear and if it does not get the right reactions (it won't) people ought to have their heads examined (they won't)."[45] At around the same time, Merton enters in his journals: "Reading a booklet on 'Cybernation.' Thought of a story, a diary of a machine, lonely and busy, still functioning very actively in a bombed-out world ten years after all the people are extinct—commenting on nothing, but brightly, busily, efficiently, in joyous and mechanical despair."[46] Merton here veers explicitly toward the motifs of science fiction, employing them as a "warrant" for alternative political thinking. His image of a machine that is perfectly calibrated, but to its own ends rather than those of species or planet survival, is perfectly cogent with the apocalyptic warnings that fill his response to Ferry. It also sets the stage for his next major book of poetry, which is concerned at length with computation and communication.

The eighty-nine cantos of Merton's 1968 book *Cables to the Ace: Familiar Liturgies of Misunderstanding* suggest that new technologies and emerging theories of communication have resulted in a crisis of language.[47] At moments, the poem evokes a familiar nervousness that new professional and technical apparatuses, technocracy and "thinking machines," have killed the capacity of words to forge bonds or profess belief. Canto 8 reads:

> Write a prayer to a computer? But first of all you have to find out how It thinks. *Does It dig prayer?* More important still, does It dig me, and father, mother, etc., etc.? How does one begin: "Oh Thou great unalarmed and humorless electric sense . . ."? Start out wrong and you give instant offense. You may find yourself shipped off to the camps in a freight car.[48]

This sudden invocation of "the camps," of mass death and Nazism, seems crushingly out of place. And yet the point is made, in a way that echoes Norbert Wiener's anguish over technological ethics in the "world of Hiroshima and Belsen." The thinking machine can be prayed to, can be addressed as a god would be addressed, not because It has allowed human beings to live lives of, say, "politics and poetry," but instead because It possesses a super-human capacity for thought. This capacity for thought, however, is a rigid system built for increased productivity, for profit, and not for ethical decision making. If a superior brain is a good enough reason to pray to something, then one may certainly pray to a computer. But in the case of a computer, one risks being sacrificed to the altar of efficiency if the prayer is not understood. If the computer were to respond, Merton therefore asks, would It respond with care—"does It dig me?"—or would It exterminate?

Most of *Cables to the Ace* is not especially concerned with thinking machines. Most of it is more concerned precisely with the cybercultural changes to human relationships in a world where thinking machines are possible. In the book's poetic eye, computers have aimed to make language into one quite definite thing: a device for communication that can either work or not work, and either invite human care, or else destroy humanity. Merton's poem removes this false choice. Poetic language instead becomes a remobilization of everyday sounds and tropes. It is a kind of noise and a kind of music, permitting a kind of protest, not against machines, but against the mechanization of thought and life. The poem begins, in its first canto:

> Edifying cables can be made musical if played and sung by full-armed societies doomed to an electric war. A heavy imperturbable beat. No indication where to stop. No messages to decode. Cables are never causes. Noises are never values. With the unending vroom vroom vroom of the guitars we will all learn a new kind of obstinacy, together with massive lessons of irony and refusal.[49]

Merton's "cables" lead to war or refusal, music or noise, but in any case do not lead to any global village or one-town world of instantaneous communication. The cables that carry messages, and secure modes of efficiency that contract time and space, also carry rhythm and noise ("an unending vroom vroom vroom") that may disrupt (through "a new kind of obstinacy") the contraction of time and space. Noises are never values—they are distortions. Like other texts of cyberculture, Merton offers a specifically, and largely science-fictional, literary distortion of his time. Like Charles Baudelaire for Walter Benjamin, the speculative form of cybercultural address is a "training in coping with stimuli"; here, they train in coping with an emerging computational order, such that an addressee might learn to "parry the shocks, no matter what their source, with his spiritual and physical self."[50]

CHAPTER 5

Revolutionary Suicide

> When reactionary forces crush us, we must move against
> these forces, even at the risk of death.
>
> —Huey P. Newton, *Revolutionary Suicide* (1973)

> dying. Long dying: long, long dying.
>
> —Joanna Russ, *We Who Are About To . . .* (1977)

The word *technology* denotes, more than it used to, emergent forms of computation and telecommunication. Where the word may once have included military and industrial machines as much as the computational components of such machines, *technology* has basically become shorthand for software and hardware and the connections between. To call oneself a Luddite in the twenty-first century seems, more or less, simply to claim that one hates smartphones and laptops and networks. As it becomes harder to talk about ways that human lives themselves have been treated as technological—as fixed capital, as instruments of labor, war, and sexual reproduction—it becomes equally hard to imagine a dismantling that might be directed inward, toward the self, as a protest against objectifying conditions. Yet the option of self-destruction, as a performative critique of the logics of power and technology, may be the most urgent expression of literary Luddism in the Long Seventies. This chapter considers a path through literary fiction of a political idea that combines machine-smashing with emancipation: the idea that an instrumentalized human body might achieve, in the risk of real self-destruction, a minimal protest and a momentary freedom.

Suicide in this chapter is figurative and not literal. Or rather, as a literary figure—in work by a range of prose writers including Joanna Russ, Paul Metcalf, Toni Morrison, and Thomas Pynchon—it is not primarily literal.

It is also a metaphor for the risk of protest against deadly social conditions. Huey P. Newton defines this largely figurative, potentially literal form of suicide in 1973: "Revolutionary suicide does not mean that I and my comrades have a death wish; it means just the opposite. We have such a strong desire to live with hope and human dignity that existence without them is impossible. When reactionary forces crush us, we must move against these forces, even at the risk of death."[1] For Newton, and for certain radical fictions, suicide is the name for a commitment to the open risk of death. It is less a refusal to live than, in Russ's phrase, a refusal to be under another's control, as "just a biological device."[2] Instead of self-murder, revolutionary suicide is machine-breaking, dismantling, practiced not on extant machines but instead on potential machines. Revolutionary suicide comes to name a figurative capacity to prevent the becoming-machine of subjugated human bodies and selves.

Just a Biological Device

Russ's 1976 novel *We Who Are About To . . .* is one of many science-fictional texts in the Long Seventies to provide speculative and experimental literary language in opposition to the existing conditions of technologized labor and life. The book's nameless protagonist—I say nameless, although Samuel R. Delany does insist that her "name *might* be Elaine"—crash-lands on an uncharted planet, along with the other passengers of her craft.[3] The passengers mill around a while, some flirt, some panic, and others make half-hearted attempts to explore, or to exploit available resources. But their activity is of no use. There is no chance for long-term survival. Several characters, succumbing to boredom in their new hopeless environs, propose a campaign to repopulate. Cheerfully, abhorrently, one of the men "has offered to donate his genetic material."[4] The protagonist is unwilling to have her body so commandeered for the survival of a species that, in any case, probably will not survive anyway. To protect herself from rape and forced pregnancy, she murders several of the other castaways, including a child, then flees to a cave. The final third of the book is a monologue: the last words of a protagonist who has slowly starved and then has poisoned herself to death.

In Rebekah Sheldon's reading of the novel, the protagonist's death is inseparable from her refusal to trade bodily autonomy for species survival. Sheldon writes: "The narrator gives up controllable predictability, and its attendant anxieties, for the complexity of a future that moves in its own course, offering some plenitude and some harm but always resisting our projections and predictions, always an epistemological void that no fence can contain nor narrative subdue."[5] *We Who Are About To . . .* describes a refusal to colonize the unknown, even at the cost of species survival, in favor of a fenceless epistemological void. Unlike other science fiction about castaways (for instance, Marion Zimmer Bradley's *Darkover Landfall*, of which *We Who Are About To . . .* is an explicit critique), Russ's novel offers no utopian vision. Just one attempt is made at social engineering, the proffered "donation" of "genetic material," and this attempt is paternalistic, misogynistic, and thwarted. Meanwhile, Russ sidelines her genre's near-universal emphasis on advanced technologies. Indeed only four machines survive the crash: a water distiller, permitting the castaways to survive for as long as they do; a gun, ultimately fired by the narrator; a meager sort of vehicle, called a "broomstick," on which the antisocial witch-hero eventually flees; and a kind of Dictaphone, called a "vocorder," on which the narration is recorded and preserved. These machines are few, just the four. They are minimal in their uses—they cannot terraform the inhospitable landscape, and neither can they provide food or offer escape. Moreover, they are flawed, mostly half-functional.

The vocorder's malfunctions are particularly salient, as it is through them that the narration becomes increasingly poetic and referential: "(This is being recorded on a pocket vocorder I always carry; the punctuation is a series of sounds not often used for words in any language: triple gutterals, spits, squeaks, pops, that kind of thing. Sounds like an insane chicken. Hence this parenthesis.)"[6] What are flaws in the vocorder become features of the novel's form. Punctuation is frequently misplaced. Sentences split in two. Misspellings. Semantic ambiguities. Parataxis. Fragmentation. Ellipsis. By the end of the novel, after the protagonist has ceased to narrate any events but her own slow passage, hallucinating and reminiscing on the floor of her cave, these formal interruptions no longer interrupt, but instead predominate, as the frequently uncommunicative voice of the novel. This is how Russ foregrounds the problematic of technology without emphasizing the technologies themselves. Even when there are only a gun, a vocorder, a broomstick, and a water distiller, these objects seem to anchor the novel's account of a future history, defining the technical capacities of a future moment. But these simple machines are displaced by a fifth machine that is more complex. Not a tool in any obvious sense, this fifth machine is the protagonist's own body, made into an instrument by social conditions. It is, if the protagonist allows it to be, nothing other than an incubation chamber for the species. Russ's protagonist finds that her body has become both tool and house for a master. It is both the space that has been commandeered for occupation and the "biological device" that might dismantle space and prevent occupation.

In response to the attempted commandeering of the protagonist's body, *We Who Are About To . . .* offers two nonequivalent Luddisms. The first Luddism is the literal Luddism that the book describes, poetically, haltingly, with "spits" and "squeaks" from the vocorder; and the second is the Luddism that the book itself is: an epistemological Luddism, a Luddism of ideas. The book narrates the protagonist's destruction of her own body, and the recording of this destruction on a stuttering machine. The protagonist's death is a Luddism that is literal, a breaking of tools, a Luddism of the body, an act that is fantastic and that destroys fantasy: the shattering within a fiction of a masculinist familialism that is all too real. The book aims to destroy the instrumentalist idea of what a woman's body is for, while disabling the generic expectations of science fiction. It aims not actually to advise suicide, but instead to unsettle the compulsions toward motherhood and continuous narration, by inverting the received wisdom that it is better to live in bondage than die in freedom. At the end of the novel, the protagonist nears death and confronts one of her hallucinations, the specter of a man she had known in her youth. To this apparition, she justifies her murderous actions:

> "You know it was self-defense." He merely looked at me. "Of course it was, of course it was!" I said. "You saw it . . . I mean running into the brush yelling Colonize, Colonize, and all that. They were going to force me to have babies . . . It was a mass-delusional system . . . and anybody who doesn't agree has to be shut up somehow because it's too terrifying.[7]

Embedded within the Luddism that the book narrates, there is the Luddism that the book is. Simultaneous with the destruction of the instrumentalized body, there is the destruction of the ideological formation that would make instruments out of certain bodies: the prevention of their use by others.

After the protagonist has murdered her shipmates and retreated to the cave, she lies on the floor, dwelling in her memories and muttering into her vocorder as her body starves. She remembers physical things: old lovers, a dentist's office. She muses on what can and can't be remembered:

> Even music . . . won't last unless you translate it into ideas . . . To mean, not be. Which the body is not good at. (For example, fucking. Why is it sometimes rememberable and sometimes not. And what do you remember? I think either a picture or an emotion but not the physical thing itself, not half an hour later, sometimes. Like a dried leaf. A dead rose. A taste that's gone . . . I can so vividly recall Marilyn.)[8]

Finally, the novel ends on a sentence fragment, as the vocorder sputters to a halt, and the narrator at last poisons herself: "I'll do it the instant way, I suppose," she says, "just to be finished with it. Get it all over, all that dying. Long dying: long, long dying, forty-two years, that's too much, and I really *wish.*"[9] The narrator is caught up in remembered events and in the conditions of their remembering. She dies.

The future is set aside and the past rises up, both what can and what cannot be remembered, then Russ's protagonist lets go at last. The novel's version of revolutionary suicide is a displacement of technology coupled to a displacement of compulsory motherhood, biological destiny, and genre convention. Russ offers no positive alternative to what she displaces, not even her protagonist's survival, but instead only an act of "dying. Long dying: long, long dying." In this way, the book stages her death as a radicalized form of Luddism, wherein a worker dares to destroy the tool that is herself. At the same time, it confines the suicidal impulse to the domain of fiction. Suicide, for Joanna Russ as for Huey P. Newton, is not a death wish. It is a movement against reactionary forces of technologization, even at the risk of death.

Acts of Lawless Outrage

Paul Metcalf's 1976 experimental novella, *The Middle Passage*, performs a similar movement, with still more explicit connection to Luddism. Metcalf's work has much in common with better known Black Mountain poets, sharing stylistic similarity with Robert Creeley and Robert Duncan; a critique of technology with Denise Levertov and Paul Goodman; and thematic and historical commitments (especially to the work of Metcalf's great-grandfather Herman Melville) with Charles Olson. Subtitled *A Triptych of Commodities*, *The Middle Passage* sees the world and the past not through the eyes of its human subjects, but instead through the translucent tissue of discarded papers. It is a collage and constellation of letters, diaries, normative histories, and balance sheets, and it concludes with a full scholarly bibliography. *The Middle Passage* divides into three brief chapters: "Ludd," about the looms, called stocking frames, that were destroyed by the disciples of Ned Ludd; "Efik," about West African slaves in transit to the Americas, suffering in the holds of ships, frequently drowning themselves; and "Orca," about the whales who, under attack from whalers, rammed the ships with their heads.

Like the world it imagines, *The Middle Passage* is partial, fragmentary, and terrifying. The words in the book are mostly quotations that Metcalf has appropriated, albeit only in bits and pieces. These are fragments of lines from historical texts, both primary accounts by participants and secondary accounts by historians, between the start of the nineteenth century and the middle of the twentieth. As disconnected as the book's chapters, the nonhuman commodities nevertheless haltingly remap a planet. The first chapter, "Ludd," tells of the Luddite assault on automated stocking frames, in words borrowed from firsthand and historical accounts:

> three hundred frames broken, thrown into the streets, a house burnt down
>   numerous acts of lawless outrage
>   military training, and the
> seizure of arms
>   clubs & sticks, swords, guns and pistols, with sledge
> and axe to wreck the frames
>   meeting secretly upon commons and
> moors[10]

These cribbed lines draw a revolutionary picture, their combination of continuity and abstraction managing two tasks simultaneously. One, they evoke the drama of a Luddite act in which tools of destruction (clubs, sticks, swords, guns, pistols, sledge, axe) are brought to bear on objects (frames, streets, house) worth destroying. Two, they apply an avant-garde poetic technique of fragmentation to the conventional prose of historical writing.

Both acts—the Luddite act that "Ludd" recounts, and the Luddite act that "Ludd" is—are conducted without a human agent. Whole sections of the text, including this section, are unpopulated by human characters. Someone must be meeting on commons and moors, and someone must be committing the lawless outrage, but the text retains only the instruments and aftereffects of these activities. And as if he were a character himself, the human figure of the author also goes missing. Metcalf's authorial voice can be heard only in the distorted reinscription and not in the creative production of written English. Because there is a bibliography at the end of the book, the reader well assumes that the words are not Metcalf's own but are instead plucked from prior sources. The force of meaning in those words is therefore not in the inventive capacity of a master-author, but instead in the novelty of the collage, the evident interest of the stories it recounts, and the uneasy fit of poetic and historical language. This force also comes from the juxtaposition of the scenes in its "triptych," as the Luddite scene gives way to the death scenes of captive slaves and of hunted whales. *The Middle Passage* is surely a book whose energies are aimed at disrupting scholarly histories of hunting, slavery, and technology—but the result is not only disruptive. Even without the presence of a unique authorial style, or of an original story, the book's collage makes something new through synthesis, an exercise in contrasts.

Its second section, "Efik," is assembled from bits of historical scholarship mixed with autobiographical writing by officers and passengers aboard slave ships. Here, quotation marks set apart certain phrases within the flow of language. Have these words been borrowed from the books in the bibliography? If so, one wonders, was more of the book in Metcalf's own voice than was first presumed?

> "it is pitiful to fee how they croud thofe poor wretches . . ."
> . . . stowed on platforms, each lying on his right side, "which is
>   considered preferable for the action of the heart"
>
>   or seated within one
>   another's legs, pressed together, backbone to breastbone
>
>   or stacked
>
> in tiers, spoon fashion, knees flexed, eyes forward, head to wind-
> ward.
>
>   or packed against the ship's
>   curved planks[11]

As with the lines from "Ludd," the lines from "Efik" perform a double role that is both historical and formal. By re-presenting historical language that is nothing but descriptive, Metcalf has allowed it to stand for itself in all its inherent violence. Yet whereas some of the passage is in quotation marks, some of it is not. And whereas the quoted passages are indeed appropriated from the historical record, they do not originate in any of the books in the bibliography. The first quoted line in this passage ("it is pitiful . . .") was written by a priest named Denis de Carli, while the second ("which is considered preferable . . .") was written by a ship captain named Theophilus Conneau. Each of these sources is easily searched and found (in this age of Google Books), but neither is footnoted or listed among Metcalf's sources. *The Middle Passage*, as a result, suddenly seems a text whose link to history is urgent but undefined. It may not in fact be lacking a poetic voice after all. Some of these lines might be Metcalf's own. But which lines are his, and which he has borrowed from his sources, and which of these sources have been cited—only research, and not mere reading, can determine.

The thematic comparison between "Ludd" and "Efik" is as tenuous as any comparison can be between early automation and the transatlantic slave trade. But as the book is just a constellation of contrasts, so the tenuous connection is the only one that can be made. It coheres around the matter of suicide, as Metcalf's text records the self-drowning of brutalized slaves. In this section of "Efik," Metcalf strings together several quotations, each in quotation marks. Each is appropriated from a different historical text, but lined up consecutively, they together tell a single story:

> "He put his head under water, but lifted his hands up; and thus went down, as if *exulting that he had got away.*"
>
> "he dived under water, and rising again at a distance from the ship, made signs which words cannot describe, expressive of his happiness in escaping. He then went down, and was seen no more."
>
> ". . . when to our great amazement above an hundred men slaves jumped overboard"
>
> ". . . they had no sooner reached the ship's side, than first one, then another, then a third, sprang up on the gunwhale, and darted into the sea"
>
> ". . . continued dancing about among the waves, yelling with all their might, what seemed to me a song of triumph"[12]

All the jubilation of the Luddite act, the destruction of stocking frames, is here repeated as the jubilation of an act of suicide. Making a disjointed paragraph out of indents and ellipses, separated by lines drawn from disparate books and papers, the passage recounts multiple scenes of bondage and mass death as if in a single narrative of emancipation and joy.

In "Ludd," humans had been subtracted from the action. The only characters who remain there are the tools of destruction and the objects to be destroyed. Then in "Efik," it is precisely those humans who have been deprived of their humanness—slaves—who become self-determining characters with motives and emotions. In making death the price of liberty and autonomy, Metcalf has mobilized a central figure of African American and Afro-Caribbean traditions in radical philosophy and political thought. From Frederick Douglass in 1846: "She leaped over the balustrades of the bridge, and . . . chose death, rather than to go back into the hands of those Christian slaveholders";[13] to William Wells Brown in 1853: "Death is freedom";[14] to C.L.R. James in 1938: "They often killed themselves, not for personal reasons, but in order to spite their owner";[15] to Orlando Patterson in 1967: "The most extreme form of passive resistance was that of suicide [by those] who preferred to take their lives rather than have the slave masters take them from them by degree"[16]—slave suicide has frequently been accepted as a singular act of resistance. Dying in freedom, it is again understood, is better than living in bondage. Metcalf's book is in this tradition, positioning slave suicide as the proper name of a self-destructive protest against instrumentalization. It is an autodismantling of a master's house (the fixed capital that imbues the slave's own body) by a tool (the same body) that becomes masterless in the process. It is a Luddite's claim to humanness that coincides with the end of that Luddite's human life.

It is also easily romanticized. For a suicide to be called noble, it must be emptied of its simultaneous qualities of abjection, loss, and desperation. Only survivors can record these deaths in terms of "happiness," "exulting," and "triumph"; but then, only survivors remain. Metcalf's book accounts for this problem through its use of citation. The text pronounces the revolutionary politics of suicide as spoken by the captains and crews of the slave ships, and as radicalized in the palimpsest of thought by Douglass, Brown, James, Patterson, and Newton (as well as others after Metcalf, notably Paul Gilroy, Ronald T. Takaki, and Achille Mbembe). Yet at the same time, *The Middle Passage* encases slave suicide in a shell of quotation marks. What go missing in the comparison, by definition, are any personal accounts by enslaved people of their suicides. Only in the voices of slaver-owners do we hear reports of slaves' suicidal acts. Only in voices of those who are already autonomous, and who limit the autonomy of others, do we hear reports on the too-brief autonomy of those who drown. The resemblance of slave suicide to the smashing of stocking frames is therefore purely formal. The possibility of a first-person narration of slave suicide is foreclosed by the mute fact of death. It is buried under the discursive back and forth between the archival material (motivated by the ideological and emotional commitments of slavers) and its re-presentation in Metcalf's book (motivated by the ideological and emotional commitments of a mostly white if largely antiracist avant-garde). In Metcalf's book, slave suicide belongs to a discourse that surrounds slavery, a discourse that is written by slave owners and slave masters, and has little to do with the lived lives of slaves.

Hortense J. Spillers, at the English Institute a decade later, in 1987, puts the problem in this way: "I want to eat the cake *and* have it. I *want* a *discursive* 'slavery,' in part, in order to 'explain' what appears to be very rich and recurrent manifestations of neo-enslavement in the very symptoms of discursive production . . . [but] I occasionally resent the spread-eagle tyranny of discursivity across the terrain of what we used to call, with impunity, 'experience.'"[17] The layered discursivity of suicide in "Efik" is where it resembles Luddism, and where the conjuncture of suicide-and-Luddism can be understood in its bearing on life after emancipation. Yet whereas a stocking frame can feel neither despair nor pain, a human being can feel both. A political value may be assigned to either act, or to both, but it can make no transit from one to the other in spite of the formal similarity. Whatever is said, it will have to fend with the same archive that Metcalf has shredded and then reassembled in "Efik."[18] What Spillers calls "experience" is what barely makes itself sensible through the book's mask of discourse, and this inaccessibility is the subject of the third "panel" of Metcalf's triptych.

The final section of the book is "Orca," which recounts the deaths of whales at the hand of harpooneers and in suicidal attacks against the hulls of whaling ships:

> ". . . the very sea is tinged red . . ."
>
> There is a low groan, growing to an echoing bellow—the lungs' last heavings through bloodclogged tubes
>
> the bloodfountain dies to a few drops, a gurgle . . .[19]

"Orca," like the book's other chapters, swells with gruesome and abject documentary images. In its unlikely contrasts, *The Middle Passage* presses back against narratives that rely on technophilia, species exceptionalism, and racism. As Hayden Carruth describes Metcalf's novel: "This is a book about the buying and selling of life for profit, a 'dark' book. . . . Yet it is all done objectively, by documentation and suggestion . . . [such that] we are left with no liberal emotions whatever, only rage and despair."[20] To understand human enslavement as comparable to the employment of automated looms and the mass killing of whales, or to understand slave suicide as comparable to machine-smashing or to a whale's desperate thrashing—the liberal political doctrines of inclusion and tolerance should be seen as a cover for historical exclusion and violence. Desperate acknowledgment of the technicity and animality of flesh (again borrowing from Spillers) should take their place.

Metcalf's textual and political Luddism interrupts the clarity of historical speech and, at the same time, interrupts the presumed relation of technology to the futures of life and work. Metcalf shares with his great-grandfather a concern with ethical and ontological forms that are inaccessible to accepted knowledge. In recollecting whale death, machine-smashing, and slave suicide, Metcalf is dragging nineteenth-century political practices into a late twentieth century for which they barely make sense. In reading *The Middle Passage*, one feels like Amasa Delano, in Melville's *Benito Cereno*, who does not recognize the slave uprising happening around him because it is anathema to his experience; or like Bartleby's employer, the attorney, for whom Bartleby's statements and condition are untranslatable because they have no place in his language of finance. Yet Metcalf's method is not philosophical, or at least he thinks it is not, writing: "I have difficulty dealing with philosophy because I view it as conclusions, or distillations. . . . Whenever I find myself reaching a conclusion, or a meaning, or a philosophic concept, I instinctively plunge it back into the day-by-day, rebury it."[21] The method, then, is one of digging into the archive for scraps of day-by-day life, quotidian and brutal, to be tossed like confetti into the face of a bewildered reader.

Is It O.K. to Be a Luddite?

Thomas Pynchon, reckoning with Luddism in 1984, wrote an op-ed in the *New York Times* titled "Is It O.K. to Be a Luddite?" He notes there that even as the conditions of technological exploitation become more and more visible, there seems less and less likelihood of rebellion. The problem, as far as Pynchon is concerned, is an increase in the power held by owners of the means of production: "Luddites today are no longer faced with human factory owners and vulnerable machines . . . [because] there is now a permanent power establishment of admirals, generals, and corporate CEOs, up against whom us average poor bastards are completely outclassed." He concludes: "We are all supposed to keep tranquil and allow it to go on, even though, because of the data revolution, it becomes every day less possible to fool any of the people any of the time." The consensus, as Pynchon sees it, is not that all machines are good. Because the generals and CEOs are not fooling anyone anymore, the consensus is rather that nothing can be done to oppose them. Unsurprisingly, the most crucial question of Pynchon's op-ed is on the role of literature: "Is there something about reading and thinking that would cause or predispose a person to turn Luddite?"[22] Pynchon does not answer this question, but clues to an answer may lie in his own work, and in that of his contemporaries, not only Russ and Metcalf but also Toni Morrison.

Pynchon's own 1973 novel *Gravity's Rainbow* aimed a precise form of Luddism at the power architectures of bureaucratic and geomilitary power, while extending this Luddism into a sort of revolutionary suicide. As Friedrich Kittler argued in 1985, the novel takes as its object the architectures of global war, but then dismantles itself, while modeling the simultaneous dismantling of those architectures: "*Gravity's Rainbow* . . . juxtaposes its own progressive decomposition to the negentropy of the military-industrial complex. The present tense alone, which Pynchon sustains throughout in contrast to the past tense employed by most novels, ensures a narrative surface that suppresses linear intersections of cause and effect."[23] The novel reproduces the experience of a present moment by eschewing the past tense employed in most fiction. Then, for Kittler, it matches the rhythms of its own sentences and story to the rhythms of the war machine. The result is a novel that "decomposes" itself, and that by metonymy decomposes (dismantles) the legitimating narrative of military institutions and technologies.

The key factor in the novel's decomposition is the progress and slow disappearance of its protagonist Tyrone Slothrop. Slothrop is a soldier whose body has been transformed quite literally into a machine of war, and whose adventures occupy all principal actions of the novel right up until the moment, many pages before the novel ends, that he disappears. The novel thus couples its own dissolution not only to the dissolution of the war machine, but also to the erasure of its protagonist, a soldier-made-machine. As Kathleen Fitzpatrick has argued, Slothrop's body has achieved a "machinic state" in which "the reader encounters a represented falling off of human agency in the face of contemporary technologies, a dehumanization that results directly in a . . . desire for a dehumanized, inanimate equilibrium."[24] While the novel includes no scene of Slothrop's death, Slothrop does go missing from the page, and so achieves inanimate equilibrium in a way that is more or less interchangeable with death. Like Russ's protagonist poisoning herself rather than become an incubating chamber, and like the shipboard suicides in Paul Metcalf's *The Middle Passage*, Slothrop vanishes from the narrative rather than continue his progress toward instrumentalization, dehumanization, and negative entropy.

When Pynchon calls for a textual procedure by which "reading and thinking would cause or predispose a person to turn Luddite," it may be that revolutionary suicide is exactly what is meant. Directed not only toward bodies as machines but also toward war machines in particular, such suicide can be found even in texts with no clear relation to technology. Toni Morrison's *Sula* is published the same year as *Gravity's Rainbow*, and its dismantling is similarly effected through a story of revolutionary suicide. The focus of Morrison's novel is on the life and relationships of a title character who lives as an exile in her own hometown. The town is an area in Medallion, Ohio, called "the Bottom," home to an unusual holiday, National Suicide Day. Begun by Shadrack, a shattered veteran of World War I, the holiday offers all residents an opportunity to commit suicide without judgment. Yet as with revolutionary suicide in the sense intended by Huey P. Newton, death is not the point. For Shadrack, rather: "It was not death or dying that frightened him, but the unexpectedness of both. In sorting it all out, he hit on the notion that if one day a year were devoted to it, everybody could get it out of the way. . . . In this manner he instituted National Suicide Day." Not a day for ending one's life, but a day on which one might die if one wanted, so that "the rest of the year would be safe and free."[25] In her review of the novel, Audre Lorde describes the holiday as a paradoxical appeal to the virtues of self-preservation: "*Sula* opens with mad Shadrack proclaiming National Suicide Day. . . . It closes with the keening voice of one Black woman's sorrow. . . . Both events are attempts to touch something which is human and redemptive—self—lying at the core of a painful and inhuman world."[26] To the extent that the world of laboring social relations is dehumanizing, National Suicide Day provides conditions for human struggle against nearly impossible technological conditions.

It is perhaps not surprising that Morrison, an editor of Huey Newton's writings and a conservator of his legacy, should have given a literary shape to the philosophical and political idea of revolutionary suicide. It may be more surprising that Morrison, whose work rarely has anything to do with technology, should have done so in Luddite terms. Yet Luddism it is when technological conditions are what have led to Shadrack's constant confrontation with the risk of death: "Shadrack had found himself in December, 1917, running with his comrades across a field in France . . . keeping close to a stream that was frozen at its edges. At one point they crossed it, and no sooner had he stepped foot on the other side than the day was adangle with shouts and explosions. Shellfire was all around him."[27] Shells and explosions, the principal technologies of world war, are literary incitement to the Luddism of revolutionary suicide. The history of technology has made it possible to leave a body "adangle" amid the noise from new and deadly machines. Consequently, that history has also made sense of Shadrack's reframing of suicide, not as a disastrous final measure but instead as a way to ward off the forms of disaster that accompany the new machines. Houston A. Baker has called it "fitting" that National Suicide Day should emerge from World War I, "capitalism's most indisputable moment of global-technological 'modernism.'"[28] For Baker, Shadrack's holiday parade is an appropriately jubilant protest against the material conditions of labor and war that have led to instrumentalized human life. "Like inversive Luddites," Baker writes of the off-season suicide parade that takes place at the novel's end, "they march against the very signs of their denial."[29]

The theory of revolutionary suicide is easily misunderstood as a radical pessimism and a promotion of literal death. Certainly this is how Jim Jones misunderstood Newton's ideas in his last words at Jonestown, Guyana, in 1978: "We got tired. We didn't commit suicide. We committed an act of revolutionary suicide protesting the conditions of an inhuman world."[30] For Jones, the rejection of global-technological modernism (to retain Baker's phrase) involves real surrender to the weariness of contemporary life. For Newton, however, Jones had missed the point, enacting a reactionary suicide in spite of himself. "I don't think my meaning ever got through to Jim Jones . . . he took the idea and used it in his own way, in a way I didn't intend at all. I can't make any sort of rational picture of what happened in Guyana."[31] Revolutionary suicide, even when it is refined by the words of literature and philosophy, can be misunderstood with horrifying results, like the nine hundred deaths in Jonestown. Newton admits as much, in a tone heavy with disappointment: "I suppose we have to be more careful about the sorts of people and organizations we get involved with in the future. I suppose we'll have to move more slowly. . . . I don't believe in utopias."[32] The theory and practice of revolutionary suicide is easily mistaken for the theory-less practice of reactionary suicide. It is easily accepted as the opposite of survival. More than anything, this what makes it impracticable. Yet in the literary practices of Morrison and Metcalf, Russ and Pynchon, revolutionary suicide is still an inversive Luddism—a routine of risk by which to march against the signs of one's denial.

CHAPTER 6

Liberation Technology

> We who are seeking ways of survival in the twentieth century begin by establishing new definitions and new fields of vision.
>
> —John Mohawk (Sotsisowah), *Our Strategy for Survival* (1978)

> Subsistence is a synonym for existence, differing only in its suggestion of endurance.
>
> —Carol Hill, *Subsistence, U.S.A.* (1973)

Technology has long been associated with the possibilities of freedom, from the Silicon Valley libertarianism that Richard Barbrook and Andy Cameron labeled the "Californian Ideology"[1] to what Seneca philosopher John Mohawk (Sotsisowah) called "liberation technology." Whereas the former way of thinking foregrounds unrestricted technological development as a synonym for human emancipation, the latter instead presses toward emancipation above all, and calls only for what tools (local, minimal, imaginative) may serve that end. Whereas the former strives toward human perfectibility and individual wealth, the latter aims to achieve the communal project of survival. What, then, are the tools of justice? With what material means, and by what regimes of making and breaking, can a community liberate itself from the regimes of power that govern it? What does technology have to do with getting free? In the Long Seventies, these questions were at the forefront of both left and right thinking. The emerging consensus, a reactionary consensus that retains its force today, was that the right kind of machine, or the right way of using it, might lead to a certain kind of liberation. In opposition to this consensus was an alternative account of freedom that looked at technology skeptically, but that nevertheless accepted the likelihood that certain advanced tools would be of use in continuing struggle.

Liberation technology, as an idea best associated with the political and intellectual activity of the Haudenosaunee Confederacy, may nevertheless be discerned first and most clearly as a figure in the work of the Osage writer and academic Carter Revard (Nompehwahthe). Writing far from the Haudenosaunee context, Revard consolidates the core ideas of liberation technology in his 1975 poem "Driving in Oklahoma." Revard writes of an experience of freedom that is lived behind the steering wheel at high speed, racing through the Oklahoma plains with windows down and radio on. The driver's freedom is vividly technological, as the poem describes the "humming rubber" and "seventy miles an hour from the windvents" of a vehicle that moves smoothly "over the quick offramp" on a route "between the gravities / of source and destination like a man / halfway to the moon."[2] Yet there is a shift halfway through the poem, as the driver catches sight of a bird that reorients his whole scene: "I'm grooving down this highway feeling / technology is freedom's other name when / —a meadowlark / comes sailing across my windshield."[3] The word "when" is the poem's hinge, between the poem's first half, recounting an experience of purely technological unboundedness, and its second half, recounting a freedom that is less steady and more vertiginous.

The joy of the drive and the humming and windows and windvents and country music—this joy veers away from the driver just as the meadowlark veers away. The bird's song, as "five notes pierce / the windroar like a flash," flips the driver's perspective as it "drops me wheeling down / my notch of cement-bottomed sky." With the swerve of the meadowlark, everything else changes too, and the order of things is reversed. The road is no longer the ground for the driver's speeding figure. It becomes the floor of the sky. When technology is freedom's other name, the earth props up the road, which is gripped by humming rubber, which is steered by a man halfway to the moon. When the bird and sky distract the driver, they are freedom's other name. Above or apart from technology, the sky is a space "defined wholly with song" through which the driver may attempt to travel. Revard's vision of technology is contradictory but it is not conflicted. In this poem, technology departs from its commonest associations. It need not be computational and need not have anything to do with automation or telecommunication or war. When technology is the other name for freedom, it is as the quotidian technology of vehicle, tires, radio, off-ramp, and speed. It is not bad. It does not require dismantling. Technology is in the experience of the verbal rush that links wind vents, window, windshield, and wind roar. At the same time, in spite of that, technology is freedom's name only until it narrowly avoids a passing meadowlark. Freedom, that moment and thereafter, becomes something that is not technological at all. It veers away from car, road, speed. It becomes an aspect of the unframed sky, and of the bird who "flies so easy, when he sings."[4] Freedom is a synonym only for the technology of everyday joy, and then just rarely and briefly. The rest of the time freedom is as elusive or as illusory as a bird, a misremembered song, or a sky. Revard's idea of freedom is an idea of unregulated life among animals and plants. It is simultaneously an idea of life that, at moments, can pass among useful (even joy-giving) machines.

This is very near to the principle of liberation technology that is articulated in the years shortly afterward by John Mohawk, in his work as an editor of *Akwesasne Notes* and as a voice of the Haudenosaunee Confederacy of Mohawk, Oneida, Onondaga, Cayuga, and Seneca peoples. In Geneva, Switzerland, in 1977, the United Nations Committee on Non-Governmental Organizations hosted the Conference on Discrimination Against Indigenous Populations in the Americas. There, the International Indian Treaty Council—a coalition of first-nations peoples of North, South, and Central America, as well as the Caribbean and Pacific Islands—won formal recognition. There too, in a series of lectures, members of that coalition laid out principles for political solidarity and self-sufficiency. Among those lectures was a four-part polemic by Haudenosaunee representatives. This polemic is now collected in a mainstay volume of Native American politics and scholarship, *Basic Call to Consciousness*, published in 1978. Mohawk theorizes liberation technology in a concluding essay called "Our Strategy for Survival," where he claims that even the most exploited peoples have enough basic material and group strategy to facilitate a transformation of shared life. A principal of spiritual and epistemological groundedness, liberation technology is also a practice of self-possession without property. As Mohawk told the UN session: "Decentralized technologies that meet the needs of the people those technologies serve will necessarily give life to a different kind of political structure, and it is safe to predict that the political structure that results will be anticolonial in nature."[5] For Mohawk, the collective use of liberation technologies can lead to unexpected kinds of joint action and creation that must, because they are collective and because they are local, lead away from domination by extracommunal forms of power.

Liberation technology is an avatar of dismantling because it reverses the hierarchies of technological development, subordinating individualist purposes for machines and privileging communal purposes, to found an anticolonial theory of human-machine interaction. Because it offers a wholesale revisioning of what machines are for, liberation technology is also legible as a literary critique of settler colonialism. Mohawk writes: "The roots of a future world that promises misery, poverty, star-vation, and chaos lie in the processes that control and destroy the locally specific cultures of the peoples of the world."[6] To oppose these pro-cesses, Mohawk's answer is not sabotage and is not merely solidarity. Nor does he hope to appropriate the technologies of control and destruction themselves. Instead, he imagines a reinvestment of affective and mate-rial energies into locally available tools and materials. These then vary by locale and historical moment: "Liberation technologies are those that meet people's needs within the parameters defined by the cultures they themselves created (or create), and which have no dependency upon the world marketplace."[7] It is indeed difficult, in a moment when this "mar-ketplace" appears totalizing, to imagine any technologies that could avoid the dictates of industrial production and international trade. Yet Mohawk imagines them, listing windmills, waterwheels, underground home con-struction, and woodlots among the liberation technologies to be found, independently of markets, in his own locality.

While such technologies must vary from place to place and time to time, Mohawk notes, they nevertheless share an opposition to the familiar ensem-ble of destructive technologies that concerned other technology critics of his moment: weaponry, industry, nuclear and coal power. What is destructive, Mohawk writes in 1979, is best neglected and forgotten in the way of ancient architectural techniques, like those that built Egyptian pyramids and druidic henges:

> One must ask why techniques of doing things could be forgotten. Per-haps it was because those techniques of doing things cost those soci-eties more than they could afford to pay. We should consider those questions when we look at technology from the perspective of the Twentieth Century. Technologies which generate centralization are the very ones which pose a danger to the Life Support Index [LSI] of our environment, just as similar (and now antiquated) technologies for the same purpose destroyed the LSI of civilizations past.[8]

Mohawk theorizes a new technological regimen rather than break the old one, but this active relinquishing of bad technologies is still a kind of disman-tling. Like most dismantlings, it includes an explicit critique of centraliza-tion: "Technologies which generate centralization are the very ones which pose a danger."[9] Indeed, contemporary digital libertarianisms themselves emphasize the decentralization of networks and distributed ledgers over more hierarchical forms of power. But Mohawk's decentralism is of a fun-damentally different sort. Whereas a libertarian might get excited by a freely flowing informational marketplace that ignores or transcends nations, liberation technology sets aside marketplaces and nations all at once.

In this way, liberation technology owes and offers much to contemporaneous thinking in philosophical anarchism, as Richard A. Falk articulates the position in 1978. Rather than the simple opposition to government or order, writes Falk, "the basic anarchist impulse is . . . toward a minimalist governing structure that encourages the full realization of human potentialities for cooperation and happiness. As such, the quest is for humane government, with a corresponding rejection of large-scale impersonal institutions that accord priority to efficiency and rely upon force."[10] Humane government, rather than no government at all, and a focus on cooperation and the realization of human potentiality: these are the principles that underlie the theory of liberation technology as well. This is not the hoped-for lawlessness of an ungoverned network. It is rather the self-rule of coordinated localities, elevated to a global attitude. Falk concludes: "For those who view our era as one of transition between the state system and some globalist sequel, the anarchist perspective becomes increasingly relevant and attractive."[11] Likewise, liberation technology emerges from a dual critique of both colonial violence and state technocracy. Rather than cede the next phase of planetary life, as if "some globalist sequel" must necessarily follow moribund nationalism, liberation technology proceeds on the assumption that something else is possible: machines to suit the community, or else no machines at all.

Liberation technology performs both a synthesis and a critique of "liberation theology" and "appropriate technology," two political imperatives of Mohawk's moment. It is a portmanteau of these two ideas: liberation theology, as a decolonial movement of religious leaders in defense of the poor, especially in Latin America; and appropriate technology, the name for a new social and cultural emphasis on the suitability of tools to tasks. E.F. Schumacher is the thinker most associated with the latter. In Schumacher's 1973 book *Small Is Beautiful*, appropriate technology is a call to use the right tool for the right job in transnational efforts to aid impoverished peoples. There is a "need for an appropriate technology," writes Schumacher, and for an appropriate criteria for judging the suitability of such technology.[12] Schumacher notes that technology is generally considered in terms of its complexity, such that a particular tool may be seen either as advanced (but hard to use) or else as indigenous (and therefore short of international standards). Dismissing these criteria, Schumacher insists that technology instead be designed only for the work that needs doing. Setting aside the goal of technical sophistication, he writes, "the equipment would be fairly simple and therefore understandable, suitable for maintenance and repair on the spot."[13] Beyond Schumacher, appropriate technology belongs properly to the discourse of Third Worldism, and to the self-rule and self-determination rather than the neocolonial rule of peoples. Tools need not make the most of recent technical innovation. They need only be usable by the human beings who are tasked to use them. Appropriate technologies might vary tremendously from community to community, and from local industry to local industry, yet they are definable by what they share in common: a readiness for use by anybody at any skill level, and an orientation toward existing work to be done.

Liberation theology, meanwhile, is a religious and political practice aimed at emancipation. In *A Theology of Liberation*, the hallmark 1971 exposition of the developing role of the Catholic church in Latin American politics, Gustavo Gutiérrez outlines the priorities of liberation theology in this way: "The Church must place itself squarely within the process of revolution, amid the violence. . . . Only a break with the unjust order and a frank commitment to a new society can make the message of love which the Christian community bears credible."[14] Without any embrace of Christian practices, the theory of liberation technology invokes the same question. What is to be the role of non-revolutionary institutions when a threat is posed to the self-determination of the communities that shelter them? The answer is that such institutions cannot remain nonrevolutionary for long, but must place themselves amid the violence. As Mohawk puts it, against the view that "the earth is simply a commodity which can be exploited" stands the certainty that "a liberation theology will develop in people a consciousness that the earth is a sacred being and that all earth on life is sacred."[15] Liberation technology, as an employment of appropriate technology and the cultivation of passionate earth-bound work and belief, is what resists the commodification of the planet and the centralization of colonial power.

Restricted and Unrestricted Technology

Liberation technology, in Mohawk's terms, is assuredly not a common view of the relationship between technology and freedom at present. In the more common view, machines of travel, computation, and telecommunication are credited both with an autonomous freedom of their own and with the capacity to produce freedom within political formations. Left thinkers have frequently sought machines that might serve as universal material conditions for the diminishment of capitalism and imperialism. Right entrepreneurs, at the same time, have used the logic of technological advancement to justify the aggressive kinds of invention and market expansion in which they were already engaged. Disagreeing on which machines would lead to what freedom perhaps, left and right technologists nevertheless agreed on one thing: there is a kind of human freedom that is innate in newly emergent technologies, especially those that involve computation. This vision is fundamentally opposed to that of liberation technology, which would pin the hope for freedom only to those devices of local make, appropriate use, or collective action.

This idealization of high technology is what Langdon Winner has called cyberlibertarianism, a perspective from which "computers and networks, it turns out, are all about freedom without limits . . . in a realm of boundless creativity, self-indulgence, profit seeking, and free-floating ego."[16] A presumption that obtains across the spectrum of cyberlibertarians is that the new machines are, or should be, available to anyone whose creativity is "boundless" enough. The most vivid example of cyberlibertarianism is Stewart Brand's famous maxim "Information wants to be free"; yet the principle persists wherever computation or automation is hinged to an increase in human freedom. The Long Seventies includes numerous examples of left cyberlibertarianism, like the signatories of the Triple Revolution manifesto in 1964, seeing cybernation as having made it possible to leave behind war, racism, and dehumanizing industrial labor; or like many science fiction writers of the late New Wave, who had begun to imagine the machines that would release humanity from racialized and gendered formations of power. And yet, as Winner demonstrates, cyberlibertarianism has a distinct rightward slant toward Silicon Valley individualism and the surveillance state, and away from practices that are noncomputational or local.

Ayn Rand, as the twentieth-century's best-known libertarian, was herself a voice for cyberlibertarian ideals. Speaking in Boston in 1970, Rand detailed her dystopic vision for life lived under the thumb of the new environmentalism that she saw as antithetical to the new technological future. Men, she said, would be forced into long commutes on public buses, as cars became illegal. Their workplace, for their sins against nature, would be destined to close. The ecologists would decimate the means for work, means for transportation, and means for entertainment. Women would be forced to choose between brutalizing days in the doomed factories on the one hand and dehumanizing domestic work on the other. Life would become boring and brief for everyone, as policy becomes increasingly antitechnological—trading viable conditions of existence for a prelapsarian fantasy of life lived before pollution and material extraction. Even, she argues, the "lowest tribe cannot survive without that alleged source of pollution: fire. . . . The ecologists are the new vultures swarming to extinguish that fire."[17] Rand's mixed metaphor here gives voice to cyberlibertarianism even as computer networks, and computers themselves, are still in their nascence.

For Rand, these machines, once found, must be used. Of her dystopic narrative, the one about men and women fighting for autonomy against the collectivists and environmentalists, she writes:

> A *restricted* technology is a contradiction in terms. . . . The purpose of the far too brief example I gave you was to prompt you to make a similar, personal inventory of what you would lose if technology were abolished—and then to give a moment's silent thanks every time you use one of the labor- and, therefore, time- and, therefore, life-saving devices created for you by technology.[18]

Unrestricted technology, invention unfettered by regulation and unlimited by concern for risk, is exactly what one would expect from the founder of objectivism and a protagonist-progenitor of libertarian philosophy. If it is true that there are reasons to protect the "labor- and, therefore, time- and, therefore, life-saving devices," from whom do they need to be protected? Rand demonstrates the danger of an all-encompassing definition of technology. Where technology is anything with a moving part, or anything that recalls the Promethean fire, or anything that can be crafted and used, there is no critique of technology that will not appear as simple misanthropy. To problematize a single technology, in Rand's logic, is to problematize all technologies. To reject or dismantle the exploitative machines of war or racism is more or less the same as rejecting or dismantling the technologies of medicine and food production. It is technology, for her, that facilitates our human futures. To undermine or unmake that technology, it follows, is to spoil the future.

Subsistence

It might be objected that practices of liberation technology, in the sense that John Mohawk intended, must demand an unnecessary deprivation. To live only with what one has made, or what one can use appropriately to the needs of one's community, seems needlessly austere, almost ascetic. Yet the point is not that those with material comforts must abandon them. Rather, the point is to oppose the kind of cyberlibertarianism that is most clearly embodied by Ayn Rand. Equally, the point is not to regulate technologies in the way Rand is convinced will be the death of us all. In a time of dire apocalyptic feeling, when it seems that we will all be killed by the machines that we cannot live without, the theory of liberation technology simply says otherwise. It offers what the anarchist feminist Maria Mies has called a "subsistence perspective." Describing an academic event she had attended, an otherwise all-male panel on population growth and climate disaster, Mies begged her colleagues to pause a moment:

> Please don't forget where we are. We are in Trier, in the midst of the ruins of what once was one of the capitals of the Roman empire. An empire whose collapse people then thought would mean the end of the world. But the world did not come to an end with the end of Rome. The plough of my father, a peasant in the Eifel, used to hit the stones of the Roman road that connected Trier with Cologne. On this road where the Roman legions had marched grass had grown, and now we grew our potatoes on that road. . . . Even the collapse of big empires does not mean the end of the world; rather, people then begin to understand what is important in life, namely our subsistence.[19]

Subsistence is not about destroying or regulating all the machines that exist, although many machines do deserve to be destroyed or regulated. Subsistence is rather about proceeding in the knowledge that the end of computation or telecommunication does not mean the end of the world. The plow is assuredly a less advanced technology than is the Roman road. But it is more appropriate to the activity of a potato farmer. It is a liberation technology and a vehicle of subsistence.

In 1973, while Maria Mies was still in the early stages of the research in India that would later lead to her subsistence perspective, there appeared in the United States a book of text and image entitled *Subsistence U.S.A.* With analysis and interviews conducted by the novelist Carol DeChellis Hill, pictures taken by the photographer Bruce Davidson, and funding from the Ford Foundation, the book offers observations, oral histories, and images of impoverished people from all over the United States. Placing high value on the words that poor people use to describe their own lives, the book also renders those lives in black and white. A Penobscot woman from Maine, a Gullah woman from the Carolinas, an ex-convict, a military veteran, a hitchhiker, a hippie: what they share is the ability to gather sufficient means to survive, even to thrive. The project of *Subsistence, U.S.A.* is to record these means and to honor the communities and individuals who have gathered them. With this project comes an array of ideological and ethical risks that Hill and Davidson do not altogether escape. When narrativized and crammed into a book, the endurance of individuals and communities can tend to appear too heroic, even epic. When it is more literary than embodied, human dignity can be subordinated to cultural or aesthetic value. The unfamiliar lives of strangers can thus be remolded into the familiar shape of fictional characters or commodities. But at the same time, the stories told in *Subsistence, U.S.A.*, like Mies's account of Trier, are effective precisely because they develop a theory of liberation technology from a subsistence perspective.

The impulse to tell particular untold human stories rests uneasily with the impulse to elevate those stories into something else, not particular but universal, not human but metahuman. In the Long Seventies, as it becomes ever clearer that enforced scarcity has won the day against shared abundance, subsistence becomes a material concern. Yet even as the means of subsistence are recorded by representational regimes of print and image, print and image are themselves recruited as among the means of subsistence. In her foreword to the book, Hill develops this idea: "Subsistence came to include principles of survival, originality, wholeness, and independence. . . . Subsistence is a synonym for existence, differing only in its suggestion of endurance."[20] Just as John Mohawk defines liberation technology apart from all the technologies that are familiarly associated with freedom in the late twentieth century, so does Carol Hill define subsistence in a way that seems to exclude all the pain and want that are generally associated with living off subsistence means. Likewise Bruce Davidson's photographs rarely show discomfort and never show abjection. The images are of faces and of useful things. The cover is a photographic still life of a sack of flour, a basket, a welder's mask, an axe, gloves, potatoes and beets, a jackknife. Other images show backpacks, train tracks, a pot to piss in, a jukebox, and a diner breakfast. These, like the plow of Maria Mies's father, are liberation technologies. They sustain the lives of communities. They do not generate wealth, but they do enable endurance, even a modicum of joy.

Freedom's Other Name

"Technology is freedom's other name when / —a meadowlark." Carter Revard's lines might ring as Luddite in the contemporary ear. Not Luddite like Audre Lorde or Langdon Winner, not even Luddite like Ned Ludd's nineteenth-century loom smashers, but rather simply a celebration of interrupted innovation. The same accusation might be leveled at Mohawk, or Mies, or Hill and Davidson, that their shared take on minimal technology provides only meager cover for a pervasive technophobia. Yet Revard's poem is not technophobic. Raising the tentative possibility that there is freedom in a fast drive across an Oklahoma plain, the arrival of a meadlowlark shows just how tentative this possibility really is. But although this poem dances on the edge between a minimalist technology (the speeding car) and no technology at all (the meadowlark), Revard's work nevertheless leaves room for computation.

Not only a practitioner and scholar of Native American literature, but a medievalist as well, Revard was an early experimenter in the digital study of language. Along with his collaborator John C. Olney, Revard is credited with "the earliest attempts to compute over a whole dictionary of substantial size."[21] The goal of this research, conducted in 1966 and 1967, was to record multiple dictionaries onto readable punch cards, and then to begin accounting for the semantic overlap of various words. This overlap, which Revard and Olney called "word-sense," would be measured quantitatively by the number and kind of words that repeat from definition to definition. As a result, synonyms might be said to share a great deal of word-sense, while terms from very different social or cultural domains would share very little word-sense. So what does a fairly early exercise in computational humanities have to do with liberation technology, or with the scene of the Oklahoma driver and the meadowlark? Indeed, why do any such a thing at all? Writing in 1968, Revard replies to this question of purpose:

> First, of course, we can save time. Second, we can get a more extensive and explicitly connected "thesaurus" of sense-descriptions and the words which can have the senses the described, for our procedures will of course turn up more synonymous cross references than are given in only the synonym-paragraphs of [the computed dictionaries,] . . . and moreover our procedures will to a great extent link sense-to-sense instead of entry-to-entry or sense-to-entry. Third, we shall turn up anomalies and inconsistencies of interest to lexicographers. And fourth, our procedures will get us a newly-structured collection of sentences, almost a "reverse lexicon" of senses, on which to carry out further processing.[22]

Revard's defense is wholly pragmatic, made on behalf of dictionary users, lexicographers, and the words themselves. In their modesty, and their refusal to claim any revolutionary effect on their discipline, Revard's goals differ greatly from many digital literary and linguistic practices that gain (and then begin to lose) fashion in the early twenty-first century.

Meanwhile, in their variety, the objectives of Revard's digital project may further illuminate his later poetic claims in "Driving in Oklahoma." If Revard's enigmatic lines ("technology is freedom's other name when / —a meadowlark") mean what they appear to mean, then they cohere well with his employment of digital tools, as these make none of the familiar cyberlibertarian claims to the emancipation of people or knowledge. In the moment when cyberlibertarianism and digital literary scholarship are simultaneously born, Revard argues only for the pragmatic and disciplinary uses of literary and linguistic computation. Emancipation belongs not to computers at all, for Revard, but only to the quotidian technologies of spiritual subsistence—the car and road, the window and radio—and even these may be swept aside with the curved flight of a meadowlark. No, the value of the computational research is, in a limited way, to understand how synonyms and metaphors work, in pursuit of that "newly-structured collection of sentences." Its value is that of modern literature, in short, rather than that of all technological modernity.

As Revard explains in 1973, in an essay that revisits the digital project: "A speaker . . . begins with a meaning and hunts for a word or phrase to embody it. I hypothesize that this amounts to a kind of lexical generation, not merely searching-of-stored-words sort of thing."[23] When Revard says that he wants his computers to link sense-to-sense, rather than entry-to-entry or sense-to-entry, he is aiming to make new sentences out of less familiar metaphors. Rather than emphasize the importance of the computer that determines the sense-to-sense correlation, it is the correlation itself that draws his attention, as a "kind of lexical generation." This is how technology can be another name for freedom. The two words, as dictionary entries, share very little semantic content. But as the end product of a process of lexical generation—drawn from the word-sense they do share—liberation—each can be read as a name for the other, if only until the startling arrival of the meadowlark.

Liberation and technology are both words that vary too widely to be linked in only one way. When they are linked in the way that Ayn Rand links them, they will invariably lead to a transnational imaginary that foregrounds national borders, only then to subsume them beneath the large-scale flow of autonomous decentralized capital. This is the right-libertarian version of nationalism. John Mohawk's theory of liberation technology meanwhile links the two terms in an anticolonial strategy, albeit one that also involves gestures toward decentralization, and toward a variety of transnational and nonnational scales of life. Liberation technology, in Mohawk's sense, is nonnational in that it rejects the givenness of other national sovereignties, particularly those that are militarily enforced. It is also national, insofar as it takes place in the separable struggles of disparate indigenous nations. Most fundamentally, however, it is transnational. With his original theorization of liberation technology, Mohawk enables a transnational coordination of struggles that are both national and nonnational. The work was written within the relatively small transnational context of the Haudenosaunee Confederacy, then read aloud in the larger transnational context of the Conference on Discrimination Against Indigenous Populations in the Americas, in

1977, where not only North American but also Central and South American delegations were represented. It is this transnational quality of liberation technology, or rather its always-already-transnational quality, that allows it to travel into contemporary contexts.

True, the objects of Mohawk's thought are hardly the only ones to have been called "liberation technology." Political sociologist Larry Diamond gives the same name to the techniques of computation and telecommunication that have been put to work in the twenty-first-century political movements of China and Malaysia as well as Tunisia, Vietnam, and Iran. Diamond writes:

> Liberation technology is any form of information and communication technology (ICT) that can expand political, social, and economic freedom. In the contemporary era, it means essentially the modern, interrelated forms of digital ICT—the computer, the Internet, the mobile phone, and countless innovative applications for them, including "new social media" such as Facebook and Twitter.[24]

Diamond's triumphant account of digital technology is similar to many in the twenty-first century, and echoes some from the previous century as well. Like Mohawk's vision, for example, Diamond's is innately transnational. But what Diamond provides is a version of cyberlibertarianism rather than an alternative to it. Leaving aside the question of how much freedom is ever actually expressed on Facebook or mobile phones, Diamond's optimism rests on the presumption that these practices of computation and communication are, or should be, universally available. Where there is no liberation technology, it is implied, there can be no freedom.

In the twenty-first century, cyberlibertarian thought can take forms that range from right wing to left. Even when it opposes reactionary attitudes toward human rights and collective endeavor, a dedication to digital revolution can nevertheless recall Rand's demand for "unregulated technology." Along these lines, Cree activist Jarrett Martineau has suggested some limits to the employment of social media in the organization of indigenous actions: "Ostensibly liberatory digital forms (tweetstorms, trending hashtags and Facebook petitions, and so on) did not compel power to respond and risked displacing forms of grounded place-based political struggle. . . . It is necessary to consider the ways in which networked action, communication and activism are inscribed within pre-existing social and power relations."[25] When activism goes online, for Martineau, it invites a range of new forms of solidarity, but these occur in a virtual space too far removed from the land and water in need of human defenders. Digital tools thus risk diluting movement-based energies into a merely phantasmatic online solidarity. Social media, in spite of the "countless innovative applications" that Larry Diamond celebrates, thus lead away from grassroots principles, and away from a development of liberation technology that would, following John Mohawk, prove "anticolonial in nature." It is this latter way of thinking that serves minoritarian cultural production and group survival, by combining ideals of decentralized power with those of autochthonous community, non-secular practices of collective belief and action, and emphasis on subsistence and appropriate tool use. Yet the odds it faces are considerable.

CHAPTER 7

Thanatopography

> *This is the world of Belsen and Hiroshima. We do not even*
> *have the choice of suppressing these new technical develop-*
> *ments. They belong to the age.*
>
> —Norbert Wiener, *Cybernetics* (1948)

> *In this earthly configuration*
> *we have, not points of light,*
> *but prominent barbs of dark.*
>
> *It's all right there on the map.*
>
> —Lawson Fusao Inada, "Concentration
> Constellation" (1989)

The world is not a network; networks are but one metaphor for the world. As I discussed in chapter 2, the media theorist Alexander R. Galloway notes the inveterate qualities of the network metaphor: when one "still believes the old myth that 'networks are enough'" to describe the world, one often also gets stuck thinking that "systems are enough to disrupt hierarchies, that networks corrode the power of the sovereign, that markets are the most natural, most democratic, and most scientifically accurate heuristic for redistributing and indeed defining knowledge."[1] The embrace of networks has often involved a conviction that there is revolution already under way, for which the digital technologies of militarism or capitalism are at least partly responsible. Contrary to such a perspective, writes Galloway, we should "acknowledge the historicity of networks . . . [and] acknowledge the special relationship between networks and the industrial infrastructure, a relationship that began in the middle of the 20th century and has become dominant."[2] This chapter follows Galloway's call to acknowledge the historicity of networks, but does so by trying to imagine a planetary technological ethics beyond networks. From a decolonial tradition of poetry and prose of the Long Seventies, there spill fragments of a language that might supplant the network metaphors emerging in the same place and time, in the United States, of Telstar, Intelstar, and Arpanet. In those days, by a seemingly common consensus, the world was reimagined as a network with the proximal ethics of a village and the connectivity of a web. But a very different genealogy of texts can be excavated apart from cosmotechnics. Thanatopography, to borrow a word from novelist Stanley Elkin, dismantles any map of the world that is drawn from the nodes of its technological triumph rather than the sites of its technological violence.

The first traces of this post-sixties perspective are etched somewhat earlier, again by Norbert Wiener. Wiener's warning about the dawning computer age remains the necessary one now: before it is a world of computation and telecommunication, this planet is defined by its machine-enabled potential for mass death. As Wiener wrote about science and technological research: "We can only hand it over into the world that exists about us, and this is the world of Belsen and Hiroshima" in which there is only slim hope that technological benefits might ever "outweigh the incidental contribution we are making to the concentration of power (which is always concentrated, by its very conditions of existence, in the hands of the most unscrupulous)."[3] Ours is a world of Belsen and Hiroshima still. It was a world of Belsen and Hiroshima in the postwar years of Wiener; it was a world of Belsen and Hiroshima in the 1960s, in the first days of the metaphor of networks; and it remains such a world. The lives and fates of extant species have changed with shifts in technology and industry, computation and telecommunication. Some of these changes have forged intercultural and interpersonal connections and a few of them have added to the health or life spans of species, yet many others have contributed equally to the acceleration toward death.

Particularly Jewish and Asian American literatures, with a solidarity that precipitated in the crucible of the sixties, foreground Belsen and Hiroshima as the paradigmatic technologies of the present. Concerned with racial and ethnic justice, during and after the movement for civil rights, during and after the protests against the Vietnam War, these literatures metaphorize the shared provenance of war machines, guns, concentration and internment camps, partitions between social movements, bombs, satellites, supercomputers, and prisons. To the extent that this metaphorization compiles another way to think about the world in its totality, it is at odds with networks.

The world of Belsen and Hiroshima, when it coincides with the world of telecommunication and computation, is a world that needs quickly to learn what and how it invents, and then to sabotage any machine that would forestall collective life. Images of such a world exist; and unlike networks, they may in fact prove to be (in Galloway's terms) corrosive to power. Post-sixties U.S. fiction and poetry is not the first literature to mark a point of contact between the Holocaust and the bomb. Characterized by an emancipatory politics and a tendency toward historical allusion, however, this late-century literature contains (more than do the literatures of the fifties and late forties) an alternative set of metaphors that opposes the ideals of technologically enabled togetherness, and calls instead for solidarity and responsibility in the face of technological dangers. Put briefly, this is the literature that, in its coemergence with the metaphor of networked and village-like connectivity, diagrams the world as a place in which to live in seriousness and mutuality with other humans and with the most dire machines. In a *thanatopography*—what Janice Mirikitani calls a fleshless shadow that is tattooed on the earth—there is a lesson to be learned about Wiener's world of Belsen and Hiroshima.

This thanatopography is a divergent account of how the world is joined together. It too belongs to the historicity of networks, inasmuch as it shows how networks were only ever one way to make sense of cosmotechnics. The celebration of networks has principally served the communications and intelligence industries whose early inventions inspired the metaphor in the first place. By contrast, as a kind of dismantling, thanatopography serves a radical memorial impulse, an ethical impulse, and the collective striving toward communion as a defense against such industries.

Holocausts. How We Compare Them

In her 1978 poem "On the Crevices of Anger," Nellie Wong gathers about her the revolutionary voices of women. Wong asks:

> And who are the poets, the writers, the painters
> who landscape the hills with words and trees?
> Who digs with hoes, what dirt loosens
> from fingernails, what smells?

And she answers:

> we the women who gather language
> into our arms, sifting, sifting
> . . . Form and content. Content and form.
> Hand in hand we move, separate, move.
> How to see with feeling,
> how to act. How the choices beckon
> if we move with love.[4]

For Wong, literary responses to sedimented social realities are possible when writers cultivate the literary soil of both form and content. Digging, sifting, tilling, smelling, speaking—all as acts of love, all as necessary activities that precede planting, and that require an intimate relation to the ground of any subsequent literary or political figuration. But what are the social realities that may be dug up and sifted in this way? Such social realities are varied and they are planetary. They are practically incomparable, but together they unsettle the earth. From sexual to colonial violence: "No earthquake but the floor moves, / it moves / below my still feet"; "rumblings in our hearts / across the seven seas."[5] What, then, is the task of literary politics but to connect the unconnectable nodes of technological misapplication and exercise of power? To Wiener's comparison of Belsen to Hiroshima, as if these two locations or events had anything to do with one another, Wong's lament can be heard as a response: "Holocausts. How we compare them."[6]

To consider the Japanese atomic blasts together with the genocide of European Jews, as if they were symmetrically comparable, is to force an incomplete comparison between historical events that have very little in common. It is also to perform a conceptual operation that has become more or less commonplace. In recent years, this operation is most often enacted in the theoretical and historical study of trauma. As Dominick LaCapra has written in a critique of that field: "Historical trauma is related to specific events, such as the Shoah or the dropping of the atom bomb on Japanese cities, [and] it is deceptive to reduce, or transfer the qualities of, one dimension of trauma to the other."[7] The manner by which the dead were killed, the duration of the conflicts, the effects on the population and environment, and the secondary effects on nonwitnesses—all of these differ, according to LaCapra. The discourse of trauma thus falls short because it links dissimilar events, thus denying each its singularity. But for all its inadequacy, the yoking of the two events does not end with LaCapra's critique any more than it begins with the field of trauma studies that is his object. Judging from Norbert Wiener's 1947 use of the phrase "Belsen and Hiroshima," the pairing is almost as old as the events themselves. Involving but irreducible to critical theories of historical trauma then, the coupling of concentration camps to atomic bombs has served several purposes: it accentuates the importance of 1945 as a principal date in the periodization of the history of modernity; it calls attention to the forms of mass violence of which humanity had become capable; it licenses, indeed demands, the institutions and injunctions of human rights that emerge at midcentury; and it fuels a critique of technology that would be based in the knowledge that camps and bombs are even more devastating than was thought possible.

In the cultural imagination, the conjunction of Belsen and Hiroshima (or another version of it, like Auschwitz and Nagasaki, the camps and the bomb, and so on) has come to suggest the experience of a limit in human memory and expression, beyond which human violence and human suffering tend to become mere abstractions. It also suggests a phantasmatic thread drawn between two diverse locales, like a string knotted to two pushpins, far from one another on a map on a schoolroom wall. Kurihara Sadako, a Japanese survivor of the first bomb, draws this thread in a 1989 poem: "What Auschwitz left behind: / turn all the world's blue skies and seas into ink / What Hiroshima and Nagasaki left behind: / a human shape burned onto stone . . . / Hiroshima, Auschwitz: we must not forget. / Nagasaki, Auschwitz: we must not forget."[8] As Kurihara's poem demonstrates, the conjunction renders an ethical demand: such events should be remembered, and remembered in relation to one another, so that they might never be repeated.

This memorial principle—never forget—has given way to internationalist institutional practices as well, like the planning of a Hiroshima-Auschwitz Peace March in 1962; the gathering of Japanese peace advocates into a Hiroshima-Auschwitz Committee in 1970; and the subsequent construction of a joint memorial, located in Hiroshima, that reads: "We should ponder over ourselves of the avarice, rage, and stupidity that are deeply infiltrated in the hearts of each and all."[9] Beginning in the same years, the thread that knots Hiroshima or Nagasaki to Auschwitz or Belsen is not only the condition of solidarity. It is, with as much force on the poetry and prose of the Long Seventies, the unifying agent in a critique of new technologies. As Raphael Sassower writes, it is a rule of the recent past that the memorial principle must be returned to repeatedly, not only for its recollection of world-historical crimes, but also for its renewal of technological ethics: "It seems that every generation of intellectuals of whatever philosophical orientation must account for Auschwitz and Hiroshima not only as instances of inhuman mass destruction but also as events whose perverse success depended to a large extent on the technoscientific community."[10] To the extent that Sassower is correct, and insofar as it is as true of novelists and poets as it is of scientists and technologists, the memorial principle is not only a caution against total war—it is also a caution against the unethical application of new tools.

It belongs to, and in part establishes, a spatial paradigm whose clearest expression is Lawson Fusao Inada's 1989 poem "Concentration Constellation," a virtual remapping of the United States, not in terms of its communication networks and internal trade routes, but instead in the distribution of internment camps. Internment camps, established in 1942 by Franklin Delano Roosevelt's Executive Order 9066, held more than a hundred thousand Japanese Americans at sites around the country. Inada maps them—Manzanar, Gila, Poston, Rohwer, Jerome, "Amache looming in the Colorado desert," Heart Mountain, the "jewel of Topaz," and the "frozen shore of Tule Lake"—in their distance from one another and in their proximity to other sites of dispossession, like an Arizona reservation and the Gulf Coast, where is found "a mess of blues."[11] Inada resolves: "In this earthly configuration, / we have, not points of light, / but prominent barbs of dark. / / It's all right there on the map."[12] As the points of orientation on a national map, Inada's poem determines, internment camps cannot be localized in an isolable time or place. They are everywhere. Indeed, the "concentration constellation" is not the exception to the rule of law, as if it were just a regretful episode that is best forgotten. It is instead a diagram of law's effectivity and the very image of an efficient national technocracy.

This constellation, a configuration of barbs of dark, is drawn together by connecting lines that may also thread other locales to one another. Inada connects the dark barb of Manzanar to the dark barb of Tule Lake partly so that he may connect each of these to Arizona's reservations and Mississippi's mess of blues. Likewise, there is a logic by which Belsen and Hiroshima serve as the first two terms in a metonymic list of terms that then extends still further. Having drawn on the familiar connection between Belsen and Hiroshima, in short, a literary or cultural text may then also associate these places with others that share in the history of technologically enabled mass incarceration and mass death. If these two dissimilar events can be connected so as to inspire a principle of unsentimental memorialization, the logic goes, then other events may be similarly connected. But the connections among these events are not the same as the connections among nodes in a satellite or computer network. As Tung-Hui Hu has noted, the purported totality of networks is premised on paranoia and exclusion: "This fantasy of the universal network has, at its core, the principle of deviance: of having a break or a rot somewhere in the network, of having circuits—or people—that are unreliable and untrustworthy."[13] That is, while networks are defined by inclusive catholicity and universal reach, they are equally defined by the inevitable failure of this reach: the "rot" at the core, or in any one deviant node, of the network. In what Lawson Fusao Inada calls a planetary configuration of barbs of dark, by contrast, everything is rot. Inada's constellation is not a global system that must keep vigil over its vulnerable nodes and sacred principles. It is a profane and unprincipled planet that is equally transformed, and made yet more vulnerable, by its human technologies. *It's all right there on the map.*

In a 1971 text that is formative for trauma theory, peace advocate and psychiatrist Robert Jay Lifton extends shared knowledge of the camps and the bomb into a critique of contemporaneous U.S. violence against Vietnamese civilians. "Like Hiroshima and Auschwitz," writes Lifton, "My Lai is a revolutionary event: its total inversion of moral standards raises fundamental questions about the institutions and national practices of the nation responsible for it."[14] In 1974, well apart from the discourse of trauma, Derek Walcott can do something similar, extending the slender thread that connects the bomb and the camps into an indictment of colonialism. Walcott writes (with echoes of Aimé Césaire) that Caribbean poets are deprived by colonialism of any shared past. All they can share is the experience of historical deprivation, such that their acts of creation are focused not through a lens of shared purpose or identity but instead through the lens of a violent past. When attention is paid only to "the paraphernalia of degradation and cruelty which we exhibit as history, not as masochism, as if the ovens of Auschwitz and Hiroshima were the temples of the race," Walcott argues, "morbidity is the inevitable result."[15] As Lifton does, Walcott invokes Auschwitz and Hiroshima as a way to understand distant episodes of mass death. For Lifton, the distant episode is the 1968 massacre in My Lai, Vietnam. For Walcott, it is a colonial violence that is older and ongoing, as well as specifically Caribbean. For both, the camps and bombs provide a way to convey a sense of historical scale. Both mobilize the camps and bombs as poetic-rhetorical figures by which to establish pedagogical conditions for instruction in other, quite separate, acts of historical violence.

Not only in the statements of public intellectuals but also in the underground press are the camps and bombs named as the first and worst in a still-growing catalogue of technological abuses. In a 1967 issue of *East Village Other*, for example, Steve Lichtgarden pens a protest against coal and nuclear power, entitled "Con Ed Crematorium." Of a proposed nuclear plant in Queens, Lichtgarden writes: "A nuclear reactor stores up more radioactivity after it has been operating for a while than that which is released by an atomic bomb"; and of the existing coal generator at the same site: "They are burning West Virginia in Ravenswood, on East 13th Street, they are burning America. . . . Yes, here the tubercular Negroes, crying spitting blood; yes, see the angel Puerto Rican baby, soot caked in her soft young eyes. No, you're not in hell. This is Auschwitz."[16] There is a crass rhetoric comparing the local effects of coal and nuclear power to the culminating disasters of World War II. But there is also a polemical truth that all these disasters are possible only because technologies were designed to capitalize on racialized principles of human disposability—principles that are repeated in the continued use of coal technology and the proposed use nuclear technology, in a place where poor people of color are most at risk. Four years later, in the anarchist paper *Fifth Estate*, the opposition is not to Consolidated Edison but instead to Control Data Corporation, a manufacturer of military supercomputers. Reporting on a march against the Michigan-based company, the unnamed author of "Not Good Germans" writes: "The U.S. drops an equivalent of more than 2½ Hiroshima's a week over Indochina. . . . The airspace is filled with every type of aircraft imaginable and much of it is coordinated by Control Data. According to a spokesman for the groups which sponsored the Rochester march, 'We didn't want to be like the citizens of Dachau who claimed they couldn't smell the bodies burning from the Nazi concentration camps in their city.'"[17]

So recognizable do these twinned figures become—camps and bombs as imaginative tools by which to invoke and understand events other than themselves—that they are increasingly conjoined with other events and not only with each other. This is how the thread stretches from pushpin to pushpin, to cover the whole techno-ethical map of the world. The effect is twofold, both to generalize the experience of world-historical violence without recourse to the clinical language of trauma, and also to specify the technological character of something unfamiliar, a human act that lies paradoxically at the limit of human understanding. Theodor Adorno had, in 1949, famously marked the heat-death of the Arnoldian cultural paradigm: "To write poetry after Auschwitz is barbaric."[18] Yet still, after Auschwitz and after Adorno, the poet Ron Welburn finds a direct inheritance of Nazi violence in the violence of prison guards that precipitated the 1971 uprising of black prisoners at Attica prison: "Attica / picked up where / dachau left its mark . . . / so they are closing down interment camps / so we all abide within circles of fire."[19] If Welburn's poem is barbaric, then it is so because it is opposed to the racist conditions of so-called civilization. It indicts the technocratic management of a prison-obsessed culture. Welburn does not write "internment," but instead writes "interment," and so invokes not only the containment and concentration of bodies, but also their burial and death. To claim homology among world-historical events is to renarrate history, construing the period between 1945 and 1971 not as a period of growth or consensus in the national society and economy, but instead as story of the survival of an insidious and specific form of violence. The "interment" of Jewish Europeans and Japanese Americans becomes the fires of Attica. Attica picked up where Dachau left its mark, without any intervening period of peace. Likewise, in the opening lines of Henry Dumas's 1960s poem "Concentration Camp Blues," historical legibility stems from its title: "I aint jokin people, I aint playin around / Wouldn't jive you people, aint playin around / They got the Indian on the reservation / got us in the ghetto town."[20] Even without mention of the bombings in Japan, Dumas's verse effectively piggybacks on the familiar metonymy so that it may show how a camp, a reservation, and a ghetto are homologous forms of containment and social control.

In 1979, the Tejano poet Ricardo Sanchez adds still more links to this metonymic chain that now includes not only Belsen, Hiroshima, Nagasaki, and Hiroshima, but also Dachau, Michigan, Queens, the Caribbean, My Lai, and Attica; and not only the technologies of bomb and camp, but also those of prison, colonial administration, reservation, "ghetto town," supercomputer, and power plant. At the other end of the Long Seventies, in a poem called "calles y callejones & memories," Sanchez writes: "we are no different / from puerto ricans, / indios, blacks, & / all the legions / of lesioned peoples / who in poverty survive, / for they, too, pray / and hope / within the streets & alleys / of their concentration camps."[21] Sanchez captures a flexible first-person plural in his "we," which likely refers either to Tejanos and Tejanas, or to Chicanos and Chicanas more broadly, but which expands well beyond either "we" to include any contained population that might share its revolutionary numbers and its historical wounds: that is, both its "legions" and its "lesions." Puerto Ricans, indigenous North Americans, African Americans, prisoners, the Caribbean colonized, the Southeast Asian brutalized, and the Jewish and Japanese victims of camps and bombs: all bear lesions and may yet emerge as an autonomous legion. These lesions, in Sanchez's terms, are both literal and figurative. They are also permanent, in a way that Giorgio Agamben names in *Remnants of Auschwitz*. In a reading of Primo Levi, Agamben concludes that the memorial principle— never forget, never repeat—must be reformulated to account for the fact that the remembered events have stubbornly refused to remain in the past: *"One cannot want Auschwitz to return for eternity, since in truth it has never ceased to take place; it is always already repeating itself."*[22] How this repetition manifests in the present for Agamben is how it manifests for Sanchez, as a permanent lesion, that is, as "the mark of the inhuman . . . the wound of non-spirit, non-human chaos atrociously consigned to its own being."[23] This permanent mark, to which neither Sanchez nor Agamben attempts to give a positive form, as if it could be figured in the poetry of one or narrated by the theory of the other, nevertheless provides the affirmative basis for a subsequent solidarity: "all the legions / of lesioned peoples."

This advocacy, an insistence both on the lasting qualities of a historical wrong and on the capacity of collective mobilization, can be said to oppose the logic of networks in two distinct ways. First, there is the metonymic linkage of parts of the world to one another by way of their shared histories of technological violence. Second, moreover, this linkage emphasizes the classed, sexed, and racialized inequalities that are normally disavowed by the utopian thinking of a networked information society. Armand Mattelart captures the problem:

> What this eschatological belief in the "information society" hides is the fact that, as the ideal of the universalism of values promoted by the great social utopias drifted into the corporate techno-utopia of globalization, the emancipatory dream of a project of world integration, characterized by the desire to abolish inequalities and injustices in the name of the imperative of social solidarity, was swept away by the cult of a project-less modernity that has submitted to a technological determinism in the guise of refounding the social bond.[24]

To Mattelart, the networked information society is the outcome of a "project-less modernity," in which technology is something that happens, rather than something that is built to determined—frequently exploitative or murderous—ends. Meanwhile, what gets "swept away" is the poetic and political project of Sanchez, Welburn, Dumas, Wong, and the others. Whereas the former can only ever speak in the language of the "corporate techno-utopia of globalization," the latter seeks the egalitarian solidarity of "legions." To follow the path of these poets is to lament, with Krysztof Ziarek, the increasingly myopic focus on "the global or planetary operations of techno-capital and world-wide telecommunicational networks"; and it is to ask, again with Ziarek, if the consequent philosophies of the world or planet might be "irrelevant as long as we pay attention to the fluid yet seemingly all-encompassing reach of modern techno-power."[25] The answer given is neither to deny the technopolitical reach of modern telecommunications nor to advocate for a better use of that reach. Rather, in the literary expression of post-sixties poetry of Belsen and Hiroshima, the answer is to understand, precisely, just what techno-power is, and how far it reaches.

It is trite to say that activist and revolutionary writers should find words to grapple with new conditions of technological and scientific responsibility. Yet it is perhaps less trite to ask that everybody seek words for the lesions and legions—in short, for the imaginative connections between and among the fatal episodes that have multiplied through failures of technological ethics and politics. These imaginative connections, and the real solidarity to which they occasionally give way, make a web among the global locales and planetary events that punctuate the space-time of history. They are a map of machines in the world quite different from the much-celebrated communicative network that begins to capture public imagination in the same years. The world of networks, whether the terrestrial network of cables or the celestial network of satellites, is a sphere crisscrossed by messages sent and received. Like the networked world, the world of Belsen and Hiroshima is a world of new and barely understood technologies. Yet unlike the world of networks, it crafts these technologies into a metonymy and a mnemonic of reactionary violence and, sometimes, of revolutionary capacity.

Jean-Luc Nancy, in *After Fukushima*, describes a rhythmic pattern that begins in the pairing of the camp and the bomb:

> What is common to both these names, Auschwitz and Hiroshima, is a crossing of limits—not the limits of morality, or of politics, or of humanity in the sense of a feeling for human dignity, but the limits of existence and of a world where humanity exists, that is, where it can risk stretching out, giving shape to meaning. The significance of these enterprises that overflow from war and crime is in fact every time a significance wholly independent of the existence of the world: the sphere of a projection of possibilities at once fantastic and technological.[26]

The world—as a rock in space, or as the sum of its human and nonhuman population—is not what is named by the metonymy that begins with "Auschwitz and Hiroshima" or "Hiroshima and Belsen." For Nancy, the bomb and camp are not the limit of human experience or of psychic coherence, but are rather the limit of the world as a technological projection and gestalt fantasy that the rock in space is anything other than a rock. The bomb and camp do not usher the human toward its phenomenal or ontological limits. They are exceptional events, to be sure. But more important, they give illusory "shape" to the world, and to ideological formations like "human dignity." By marking the edges of a sensible world, they create a world, and do so by disintegrating the other more optimistic or more technophilic versions of the world. For Nancy, as a result, "Auschwitz and Hiroshima have become names . . . that name only a kind of de-nomination—of defiguration, decomposition."[27] In Galloway's phrase, they are "corrosive to power" at a degree that the network can only dream, or be dreamed, of being.[28] Through the metonymy of such names are linked the coordinates of a decomposing (in both transitive and intransitive senses of that word) web. This web becomes, in short, the way the world comes together in having come apart, the way it comes to life on a pile of bones, the way it transforms through technological invention that is wholly continuous with technological-enabled disaster.

Thanatopography

In Stanley Elkin's 1977 short story "The Conventional Wisdom," this metonymic web is a "thanatopography": a map of the terrain of death. Elkin's
story is in three parts: the first tells of the daily life and sudden murder of
a Jewish liquor-store owner in Minneapolis; the second tells how that shop-
keeper, Ellerbee, is shown the glories of Heaven by an angel of death; and
the third shows Ellerbee's rejection from Heaven and his arrival in Hell.
Giving Ellerbee a grand tour of the afterlife, the angel of death confesses:
"I talk too much. I sound like a cabbie with an out-of-town fare. . . . Showing off death like a booster. Thanatopography."[29] Thanatopography is not
thanatography. It is not the writing of death (*thanatos* + *graphy*) but rather
the writing of the place of death (*thanatos* + *topos* + *graphy*). It is the map of
death, a guide through the technologically death-bounded world, showing
off death like a booster. A transformation of the world into something other
than itself, thanatopography is a label the size of the earth. Yet although a
thanatopography is the proper name of the earth, standing in for the earth,
it is nevertheless a material thing. As Ellerbee shouts while standing in hell-
fire with the damned, in a final confrontation with a vindictive God (Ellerbee has received infernal punishment for having worked on the Sabbath,
and for thinking that paradise looks like a theme park): "Maybe whatever
is is right, and maybe whatever is is right isn't, but I've been around now,
walking up and down in it, and *every*thing is true. There is nothing that is
not true."[30] What has been imagined, Ellerbee cries, is as true as anything
else. There is nothing that is not true. The transglobal narrations of death
and afterlife are not mere palliatives and warnings. They are as real as anything else.

At last, amid the flames, Ellerbee pleads with God: "I am praying to You
now in all humility, asking Your forgiveness and to grant one prayer. . . . To
kill us, to end Hell, to close the camp."[31] Ellerbee thus makes two professions
of feeling that link together to define Elkin's thanatopography. First Ellerbee
tells God that there is nothing that is not true; then he prays that God end
Hell and close the camp. Everything is true, Hell is true, the camp is Hell, the
camp is true. The narration ties an affective-epistemological knot around a
historical anchoring point, the camp, which does not merely resemble Hell,
but is Hell. Thanatopography is therefore not the imagined map of the terrain of death after life, but is in truth the map of life as a terrain of death.
The camp is a privileged site on that map. Thanks to postwar reconstruction
efforts in central and Eastern Europe, no camps remain in their wrecked and
liberated final state. Any that remain are museums. Yet though camps are

mostly absent, half erased, they persist as material evidence of the need for an injunction toward technological responsibility. As Hell, the camp remains a synonym in the existing world for a kind of death that must at all costs be avoided. Thanatopography is an infernal injunction that can issue from the world of the living because Hell, though it may not be here or now, is nevertheless true.

Walter Abish, another Jewish American fiction writer, adds depth to Elkin's image. The very earth, in Abish's story "The English Garden," is made up of technological and historical layers. A clean new surface layer above barely conceals the substratum of ash left by technological mal-feasance of the past. Set in a German city called Brumholdstein, built on the rubble of a concentration camp called Durst, the story begins with (and takes its title from) an epigraph from John Ashbery's 1956 book *Three Poems*: "Remnants of the old atrocity subsist, but they are converted into ingenious shifts in scenery, a sort of 'English Garden' effect."[32] The story is, in large part, an exploration of this epigraph. The Shoah is over but not done. The remnants of the old atrocity remain a foundational yet sup-pressed aspect of German planning and expertise. A first-person narration details the arrival of a foreign writer in Germany. This writer, the narra-tor, plans to interview another writer who lives in Brumholdstein, but the interview is more of a pretext than a purpose for his visit. Chiefly, it seems, he is concerned with assessing German social conditions three decades after the closing of the camps. Even as Brumholdstein is built on the sur-face of the well-concealed ashes of Durst, so do Germans "have a deep and abiding belief in perfection" of surfaces, like those of "a well-designed city, an attractive park with picnic tables and large shade trees, or a power-ful motorcar engine, a formica-topped counter, or the white enamel of a coffee pot."[33] The idea of a perfect exterior is thus two things at once: an aesthetic ideal of good design, and also an ethno-political characteristic of a bureaucratic nation.

The German taste for perfection, however, depends not only on what is concealed by the flawless shell of formica or enamel, but also on what is revealed by it. Durst, says Abish's narrator, "took second place to the more notorious camps such as Dachau, Auschwitz, and Treblinka" such that "after giving it some thought, the community decided that the for-mer concentration camp was not worth keeping as a monument." There is little to be gained from conservation and much to be gained from moving beyond the disaster, when "for the price of rebuilding and maintaining the Durst concentration camp they could build 2,500 apartment units."[34] The disaster thus becomes part of the ground, an element in the soil, on which is built the economical German city. Brumholdstein is named after its famed resident philosopher Ernst Brumhold. Brumhold is quite clearly modeled on Martin Heidegger, whose thought is comically reduced by Abish's narrator: "By now he's an old man. He no longer lectures to young Germans. He spends his days thinking and writing . . . writing about why humans think, or fail to think, or try to think, or flee from thought, thereby compelling everyone who reads or tries to read his rather difficult book to think about whether or not they are really thinking or just pretending to think."[35] Heidegger, still alive at the time of the story's first publication in *Fiction International* in 1975 but dead by its republication in Abish's 1977 collection *In the Future Perfect*, is here reduced to a mere puzzle maker. Brumholdstein's residents, like Brumhold's readers, think hard on the epistemological problem that Heidegger, in late work, called *Denken*. Yet with all their thinking, and their thinking about thinking and not thinking, they fail to attend to the remnants of old atrocity. This atrocity, Durst, resides below the surface of their streets and parks.

What Elkin calls "thanotopography" and Walter Abish call "surface," Harry Mathews and Georges Perec call a "secret topology." Their coauthored story, "Roussel and Venice: Outline of a Melancholic Geography," is a map of a terrain invisibly marked by death and absence. Ostensibly a serious scholarly essay about the impact of travel on the experimental fin-de-siècle prose of Raymond Roussel, Perec and Mathews's story is in fact a fiction, and a subtle satire of practices in scholarly writing and citation. Telling the story of a fictional journey to Venice by Raymond Roussel, Perec and Mathews recall both Elkin and Abish. Like Abish's Brumholdstein, a pristine surface built on ashes, this "Venice is complete and isolated: a whole world, a planet" that seems somehow to both have and lack a history; and like Elkin's Heaven and Hell, where everything is true, this "Venice is pure theater; it is a trompe l'oeil in the context of illusion itself, which can therefore be taken literally."[36] The ground of this Venice is phantasmatic, self-contained, and true; and nevertheless it is suited for foot travel: "a 'secret topology'; Venice is a city made for walking, where one is never quite sure which way is north or south, where one never knows quite how far it is from one point to another, where the link between two points is a question of continuity and/or discontinuity of surface."[37] It is an impossible space, like a cross-cap or Klein bottle, that occupies three dimensions but is without depth.

A critical commonplace holds that every work of Perec is a meditation on absence—specifically on the absence of his mother, lost to Auschwitz. Perec's *La disparition*, from which is famously subtracted the letter *e*, is in this way said to be missing Cyrla Perec. The letter that is not there is said to stand in for the author's mother who is also not there—or, in Warren F. Motte's words: "The *e* . . . figures the mother and, by extension, the parents; its absence recapitulates their absence."[38] Much more relevant to the Roussel "essay," however, is the diagnosis by one of Perec's famous psychoanalysts, Jean-Bertrand Pontalis. In his case studies, Pontalis writes: "[His] mother had disappeared into a gas chamber. Beneath all those empty rooms which he was never done with filling, there was that room. Beneath all those names, that which has no name. Beneath all those relics, a lost mother who had left not the slightest trace."[39] Without judging the clinical utility of this description, without ignoring Perec's coauthor or giving in to the psychobiographical impulse, one might nevertheless accept that Pontalis is on to something. "Roussel and Venice," at least, has not marked the absence of Auschwitz by making elements go missing. Quite the opposite, the horror of this story is that there is no depth and therefore nothing missing. The room below the surface is a fantasy of a crypt. Put another way, the surface appears as an abundance of signifiers but it has nothing beneath, only the imputation or presumption that there must be more than surface.

Pontalis says of Perec that all his rooms are filled with words, minus the gas chamber in which his mother was killed. But the gas chamber is no longer there. It was never there. It was only ever the legible figure of a loss, giving way to an infinite regress of meaning. The truth of the secret topology is not only that its precipitating cause can be found neither at nor below the surface; it is also that surface and depth (cause and effect, Hell and Heaven, Durst and Brumholdstein) are only effects of one another. Auschwitz and its efficient technologies are both everywhere and nowhere found. This is but one way to understand a story that, unlike the stories of Elkin and Abish, resists assimilation to the genres of post-Holocaust literature. Yet it goes some length toward explaining Mathews and Perec's subtitle: "Outline of a Melancholic Geography." Partly parodic of academic critical theory, the phrase "melancholic geography" partly also gives this topology its theoretical complexity. It is a geography that has never relinquished the remnants of the old atrocity. If Elkin's truth is that Hell is navigable and present, not hidden; and if Abish's thinking is that the camps are constant components of contemporary reality, visible in their absence from the surface of the English garden; then the secret topology of Mathews and Perec holds that a surface needs no depth in order to incorporate its losses. The technological threat of the Shoah is never entirely missing from expression, on this composite view, but is always formative of it.

In developing their topological theory, Mathews and Perec quote an article by a certain O. Pferdli entitled "Autres Images de Mélancolie," from a psychoanalytic journal whose title is abbreviated *Nouv. Rev. fr. psych.*[40] In fact, no such author or journal exists; and in fact, the essay's theoretical material is (partly quoted, partly paraphrased) from "Mourning *or* Melancholia: Introjection *versus* Incorporation," an article published by the French-Hungarian analysts Nicolas Abraham and Mária Török in *Nouvelle Revue de Psychanalyse* in 1972. When Mathews and Perec cite the fictional Pferdli (i.e., the real Abraham and Török) in "Roussel and Venice," they are chiefly concerned with the aim of incorporation "to avoid certain unbearable words . . . that evoke the loss of objects of love"; yet they also insist that "the object whose loss is thus denied continues to exist, as a body of actions, feelings, and inexpressible words in a secret topological system."[41] As in Pontalis's account of Perec, something seems to have gone missing from the secret topology of Venice. Yet the lost object retains a kind of ancillary presence in the overabundance of verbal expression at the ostensible surface of "actions, feelings, and inexpressible words" that are nevertheless made manifest.

This is the world of Belsen and Hiroshima, where the very skin of the planet is a ledger of technological misapplications; where everything is true, even what should never be imagined; where there is no depth to experience, but only the phantasmatic secrets of a complex topology. The metonymy of Belsen-and-Hiroshima-and- is a merely formal trajectory. It links historical and geographic sites, and by linking them, it issues the political demand that they never be repeated. The event of this linkage, to use Shoshana Felman's phrase, is an event of teaching: "In the era of the Holocaust, of Hiroshima, of Vietnam—in the age of testimony, teaching, I would venture to suggest, must in turn *testify*, make something *happen*, and not just transmit a passive knowledge, pass on information that is preconceived, substantiated, believed to be known in advance, misguidedly believed, that is, to be (exclusively) a *given*."[42] What Felman calls a "given" is the cosmos in the cosmotechnics. It is what is assumed to precede any subsequent event or connection, and what appears, since the midsixties, to be a network. The rejection of this given, as a given, is the thanatopographic remapping of the place where technologies, whether for good or ill, are still invented and used whether or not they ever connect anything. The unsettling metaphors of surface and memory and obligation thus culminate in a literary project that in turn can be said, with Felman, to be a pedagogical project. Through them, the knowledges of technological history and technological responsibility are not transmitted but are transformed.

Human Shape Burned onto Stone

In the literary politics of catastrophe there repeats not only a patterned metonymic rhythm, but also a repertoire of particular images, of familiar if terrible figures. One of these figures is the image of a concentration camp that is buried just below the placid surface of late twentieth-century life, or that persists in it, giving the lie to the illusion of depth. Another is the figure of a human body whose black silhouette was burned into the earth by the atomic blast of Hiroshima or Nagasaki—what Kurihara Sadako called "a human shape burned onto stone." Analyzing the appearance of this figure in poems by Galway Kinnell and William Dickey, Daniel Tiffany has called this scorched shape a "material shadow" and a "chiasmus of bodies and pictures"; a "body whose radiant and volatile substance is disclosed only by a nuclear event" and a "body disappearing in the catastrophic medium of the atom."[43] This darkened patch of earth is the profile of a human being, disclosed by the bomb, as a synthesis of bomb, body, heat, and light. It is indexical evidence that (to borrow from Roland Barthes) these things have been there. It does not become part of the earth in the exact way as does the English garden, the thanatopography, or the secret topology. But like those melancholic geographies of the Shoah, the material shadow of an atomized silhouette is one of the traits that define the world of Belsen and Hiroshima.

This material shadow is the central image of the fifth canto of Alexander Kuo's 1974 poem "New Letters from Hiroshima." Kuo's poem tells of the bombing from the moment the *Enola Gay* takes off until the second bombing three days later in Nagasaki. Canto 5 tells of the first bomb and its effect on bodies, while also suggesting it as a version or embodiment of U.S. racism. The canto describes the Hiroshima blast: "First a flash / A noiseless blast / then white heat / A great wind / A fire wind"; and then, in terms that are horrible but almost clinical, describes the effect: "That fused quartz crystals in granite blocks / That roasted exposed children only their shadows remained / That welded women into asphalt and stone forever."[44] There are few words for the violence of these agrammatical lines, partly because they are tragic and partly because they are macabre, but mostly because the bomb itself has gone missing. Even the flash, the blast, the heat, and the wind are only effects of the bomb. The bomb itself never enters the scene, in Kuo's recounting of events. Burn marks and geological metamorphosis are all that is left to show that something unseen has happened.

As media theorist Akira Mizuta Lippitt has written: "A blinding flash vaporized entire bodies, leaving behind only *shadow* traces. The initial destruction was followed by waves of invisible radiation. . . . What began as a spectacular attack ended as a form of violent invisibility."[45] What is most invisible is the radiation that seeps into the earth and walls, and into the bodies of survivors. But the bodies of some dead, once they have become black shadows on the ground, are not much more visible than that. Their blackness, for Kuo, is the point: "Black letters of newsprint burned right out of the white paper a mile and a half away . . . // The white was untouched, unscorched, only slightly singed // And the dark, the darker, the darker, always the darker burned into enemies forever."[46] Whiteness is untouched by the blast, while blackness burns right away. The nonwhite bodies of the civilian dead are fully disposable, except for the permanent shadow that remains. And then "the darker, the darker" is "burned into enemies." The adjective "darker" is employed as a noun, and then as both the subject and the object of the verb "burns." In this way, "the darker" names those whose bodies are darkened by the blast, just as it names those peoples of the world whom U.S. imperialism makes into enemies.

Janice Mirikitani, the poet and San Francisco activist, published a poem in 1979 called "August 6." Commemorating the devastations of the bomb, Mirikitani's poem also ironizes the language employed in the United States to legitimate the attack: "Hiroshima, / your children / still dying // and they said / it saved many lives // the great white heat / that shook flesh from bone / melted bone / to dust // and they said / it was merciful."[47] For Mirikitani, contrapuntal voices are required for an understanding of the events of its title date, the day the bomb fell on Hiroshima. It is not enough to condemn the bombing as a past wrong, the poem implies with these counterpoints, when bombing continues to be justified by a clichéd nationalism that must equally be condemned. Hiroshima's children are still dying, their death saves lives; the great white heat (a phrase that buzzes with its variation: great white hope) has shaken and melted bone, it is an act of mercy. The bomb is not only the site of a catastrophe, in these lines, but must also be understood as the site of this ideological contradiction. The contradiction continues: "Sister they found / tattooed to the ground / a fleshless shadow on Hiroshima soil // and they said / Nagasaki."[48] The rhyme of "found" and "ground" is unsettling. It is too trite, almost quaint, almost singsong. And the importance of the counterpoint—"and they said / Nagasaki"—is unclear. Do "they" say Nagasaki because the event of the Hiroshima bomb, however singular, may be diminished because it is not unique? Do "they" say Nagasaki because the word shows just enough knowledge of historical context that they never have to deal with the "fleshless shadow" or the soil on which it is tattooed?

The fleshless shadow is further reframed by the Beat-movement poet Nanao Sakaki, also in 1979. Sakaki's poem "Memorandum" is a disordered timeline, a journal that chronicles personal journeys and lost friends on various dates not in sequence, mostly from the 1940s and 1970s. One entry reads:

> 1946. Hiroshima. There,
> one year after the bombing
> I searched for
> one of my missing friends.
> As a substitute for him
> I found a shadow man.
> The Atomic Ray instantly
> disintegrated his whole body,
> all but his shadow alive
> on a concrete wall.[49]

This shadow man haunts the other dated entries in the poem. Like the 1946 entries, the entries of the 1970s all point back to the events of 1945. In 1973, the speaker visits a priest who had held Mass for pilots of the *Enola Gay*. In 1970, he visits White Sands National Monument, where "Dr. Albert Einstein, / Government officials and the Pentagon / they all watched / the mushroom-shaped cloud" of the first nuclear tests at the Trinity Site. Sakaki's poem ends in 1979, in the Chihuahua desert, where an "extincting species" of whooping crane will be replaced by a load of "ever-existing Nuclear Waste."[50] The shadow man, the material shadow, the fleshless shadow—like the English garden or the secret topology, this image is both a quality of the ground, in a world of Belsen and Hiroshima, and a figure that links to other figures in the historical progression. Just as the shadow man is a "substitute" for the lost friend, the waste from nuclear technology is a substitute for a New Mexican whooping crane.

This is how the metonymy works. The specificity of one technological disaster gives way to a series of other, equally specific and not-at-all-interchangeable technological disasters. Likewise, the Laguna poet Paula Gunn Allen cites Hiroshima to extend an understanding of the bomb toward an understanding of war crimes in Vietnam and the experience of Pueblo war veterans in White Sands, New Mexico. Yet in so doing, she too makes use of the image of a fleshless and material shadow. In the 1978 poem "In a Tavern at White Sands," Allen writes: "Black leather anger / glistens in your eyes / napalm waiting to be fired . . . / Your face / has looked empty / on crushed shells and agate / crystals in the sand . . . / the shadow in Hiroshima / of an atomized old man."[51] Allen employs the same image that also appears in Kurihara, Sakaki, Kuo, and Mirikitani: the human shape that was scorched by the explosion into the ground in Hiroshima. But for Allen, the emptiness of that human shape is akin to the emptiness experienced by one who has seen both the napalm falling from the sky and also the white sands (glinting and empty like "crushed shells and agate") of an expropriated ancestral home that became a nuclear test site.

So again the string of associations is set in motion. Where the camps and bomb share a capacity to mark the earth with death—transforming the ground into a skin on which the many dead are tattooed, burying mass death as a remnant in the soil of a garden—some similar capacity can be seen to inhere in colonies, battlefields, "interment" camps, desperate neighborhoods, Attica, the Chihuahua desert, and any other site of failed technological ethics. Some connection may also extend to other forms of mass death, most notably to the nonhuman plant and animal species whose deaths are often taken as synecdochic for the death of the planet itself. To return to the poem with which I began, by Lawson Fusao Inada, this series of connections and not-quite resemblances should suffice as a warning about technological invention:

> Now regard what sort of shape
> this constellation takes.
> It sits there like a jagged scar,
> massive, on the massive landscape.
> It lies there like the rusted wire
> of a twisted and remembered fence.[52]

What Inada calls a constellation of "barbs of dark" encircles the globe in the shape of a jagged scar, a wound. It gathers up a virtual image of a planet. It is an image that is perhaps capable of corroding the power still wielded by a figurative network that was born at the same time and from the same midsixties tumult of invention and destruction.

One twin is the network. It purports to need mending, completion, expansion, until at last it might speak for and to every human. The other twin, the poetic configuration of barbs of dark, is born moments later. It distances itself from the first twin, labeling it a liar, a shill, and an advocate of inegalitarian techno-utopia, while claiming nothing for itself but the permanence of a past (a lesion that might give way to legions) that is built into every computational and communicational apparatus, and into the techno-global apparatus of the world. While the image of the network remains fairly static, expanding in scope but rarely varying in its basic qualities, the twenty-first century has never ceased to transform the map of planetary death, adding to the iterative chain that begins "Belsen and Hiroshima and . . . ," and extending from My Lai and Attica and Manzanar to sites well beyond. In this thanatopography, which is a map of demands for new regimens for technological responsibility, one could add any number of contemporary technologies: guns dispersed into the hands of police or children; the camps that hold migrants and refugees; the fast mechanisms of mass incarceration and deportation; the surveillance systems that continue to track individual data and metadata; drones.

Conclusion

American Carnage and Technologies of Tomorrow

I began writing this book three years ago, and a lot has happened in that time. In early 2016, it seemed certain that the United States was about to enter a new phase of neoliberal governance. With Hillary Clinton as its chief executive, the U.S. would commit itself even more to markets and market logics—austerity, efficiency, scarcity, productivity, precarity, fungibility, flow—than it ever had, there seemed little doubt. And yet this happened instead. True, neoliberalism does proceed apace in a world of game-show presidencies, as industries of finance and high technology continue to consolidate their central role in the national culture and economy. But now alongside neoliberalism, surprisingly to many, there also rises a familiar form of neofascism. Donald Trump may have snatched the crown, but he did not create this neofascism. He did not invent spaces for racist, sexist, or nativist violence. He did, as proxy for a movement, widen existing spaces both online and off-line, embolden their populations, and legitimate the state and nonstate agents who have long been tasked with their defense. What happens in these spaces is a pastiche of existing legal and extralegal violences, renewed in the terrifying present on behalf of an imagined past. Yet the speed of the renewal, and the variations on the pastiche, are new. While certainly neofascism is fueled by Trumpian politics, it is fed equally by arrangements of power and technology that long predate this presidency.

Tombstones across the Landscape

The 2017 presidential inauguration speech, read by Trump from a script written by Steve Bannon and Stephen Miller, was a triumph of dystopian technopolitics. The two best-remembered sentences of the speech are these:

> Mothers and children trapped in poverty in our inner cities; rusted-out factories scattered like tombstones across the landscape of our nation; an education system, flush with cash, but which leaves our young and beautiful students deprived of knowledge; and the crime and gangs and drugs that have stolen too many lives and robbed our country of so much unrealized potential. This American carnage stops right here and stops right now.[1]

Trump's speech characterizes the terrain of the United States as a graveyard of failed institutions. The "tombstones" are "rusted-out factories" but also "inner cities," schools, and streets that still bleed, even as decades of welfare liberalism have left their managers "flush with cash." It is a thanatopography, albeit one that is constructed neither for collective memorialization nor for a radical fostering of group life. Instead it is constructed for the suppression of life. The presumptively black and Latinx populations of the "inner cities" are, in this florid nightmare of racist prose, as responsible for the "gangs and drugs" as for the fatherless children. The presumptively white employees of "rusted-out factories" are equally their victims. Carnage is the sum.

"American carnage" is not merely the bloody urban violence that is imputed by Trump, Bannon, and Miller to the gangs and drugs and fathers of the racialized city. It is also the public death of public works, as the speech effectively announces that there will be further defunding of schools and of utilities that have been deemed "flush" but wasteful. It is also the death of industry, where functional factories, fully staffed with precarious laborers, are considered an absolute good; and where such factories are abandoned, "rusted out," by trade policies that drive their owners to find labor in climes that are still less regulated, still more precarious. It is also the death of dog-whistle politics, neither through the elimination of racism nor through its replacement by an even more refined form of nudge-and-wink racism, but rather in a presidential inauguration speech that glances at the expiration date on canned forms of racist explicitness and decides, sure, they're still good.

Later in the inauguration speech, Trump explains which configuration of machines and institutions must replace the American carnage. He bellows: "We stand at the birth of a new millennium, ready to unlock the mysteries of space, to free the Earth from the miseries of disease, and to harness the energies, industries and technologies of tomorrow. A new national pride will stir our souls, lift our sights, and heal our divisions."[2] These two sentences, while they receive less attention than the other two, offer as much insight into the politics of the regime that they inaugurate. Schools are to be replaced by pride. Gangs, drugs, and single motherhood are to be replaced by a healing of disease and division. A racialized symbolic is to be replaced by a postdivision imaginary. Rusted-out factories are to be replaced by the industries and technologies of tomorrow. Science and technology, in the rhetoric of the speech, serve a notably different purpose from the one they had served in Trump's campaign. Then, medicine had been designated as a luxury commodity to be afforded only by those with money for private insurance. Then, objective claims of the environmental sciences had been reduced to subjective matters of differing opinion, and technology was to be made an exclusive priority of schooling. Now, with the inauguration speech, technology replaces schooling and licenses a new reason for national unity, while medicine, space, energy, and industry become interchangeable devices of this unified national future. By endorsing "technologies of tomorrow" as a quick fix for social schisms, rather than as the applied research of task-oriented scientists, Trump's inauguration speech shows its solutionist hand. The promised technologies, as is so often the case, have little to do with actual practices of invention and use, and everything to do with power consolidation out of control.

It should increasingly be acknowledged that present forms of fascism owe much to current attitudes toward technology. At the same time, it might equally be felt that changed attitudes toward technology will lead to changed attitudes toward power. As bigotry stretches across the Internet (less and less confined to comment threads on 4chan and YouTube) and resumes its accustomed place in the public sphere, the chief executive gets too much credit. Trump is the expression, not the agent, of a brutal technopolitics. Blame must equally be borne by the multiplying means and media of technopolitical assembly. All is perhaps not lost, however, if the conceptual tools of early cyberculture retain any of their protestatory force. In this book, I argued that the literary archive of the United States, especially from the decade and a half before 1980, can ground a skeptical perspective on present connections between technology, communication, and globalization. By breaking apart certain machines, we can learn to use them better, or never use them again. By dissecting certain technocentric cultural logics, we can likewise challenge or reject them. This perspective stands opposed to a common-sense attitude toward advanced technology that now obtains in many quarters, threaded between the extremes of technophilia and technophobia. When we hew to this common-sense attitude, we accept that some skepticism must be turned toward the triumphal narratives of technological globalization, but at the same time we caution against mere Luddism. With a common-sense attitude, we acknowledge that some modern machines are antisocial and exploitative; but then again, we also insist that we not throw out the innovative baby with the retrogressive bathwater. We are circumspect about our ballooning hours of screen time; but then again, we remain on guard lest some zealot start smashing our computers. Progress may be a myth, and its prophets may be motivated more by the profit than the prophecy. Then again, without those prophets, we wouldn't have any phones at all, and you don't want someone to smash your phone, do you? Do you?

The weak position, this common-sense attitude toward technology, floods social media in Twitter threads, in long Facebook notes, and in so-called think pieces that do very little thinking. The vagueness about what counts as technological, the cry that an attack on one machine is an attack on all machines, the presumption that there is any such thing as a singular social good or that there is any such thing as a unitary world in which there is any such thing as a right way to use machines—all these statements are enabled by common sense. Against common sense, it remains possible to instantiate what Antonio Gramsci called "good sense." For Gramsci, common-sense attitudes must be fought, yet they also tend to reveal clues as to their undoing. Indeed, for Gramsci, even though "criticizing and going beyond common sense . . . can only be conceived in a polemical form and in the form of a perpetual struggle," nevertheless "the starting point must always be that common sense which is the spontaneous philosophy of the multitude."[3] Gramsci refers to this starting point as good sense: "The healthy nucleus that exists in 'common sense,' the part of it which can be called 'good sense' and which deserves to be made more unitary and coherent."[4] So then, the starting point of the present book is also its end point, whatever good sense it may have. The ethical formula of Alice Mary Hilton—machines for HUMAN BEINGS or human beings for THE MACHINE—and the political warning of Audre Lorde—the master's tools will never dismantle the master's house— find their synthesis in a formula that may be just as sensible and pragmatic: not everything should be dismantled, but many things should be and some things must be, even if we don't know where to begin.

In the advancing twenty-first century, globalism looks to be the watchword of neoliberals and multinational capitalists, while antiglobalism has become the project of neofascists and isolationists. Teletechnology, meanwhile, is accepted as a social good by all but the most militantly unplugged.

Apart from globalism and antiglobalism, then, aside from the neoliberals' or neofascists' own lexicon and aside from technophilia and technophobia, there seem to be fewer and fewer words for criticizing the roles of technology and communication in the violent shaping and reshaping of the world. Technophiles, especially in industry and government, still celebrate the machines that enable speed in production, simplicity in travel, losslessness in data storage and transmission. Skeptics, both in and outside those sectors, warn that there are threats concealed in the apparent advantages. Meanwhile, theorists of media and technology frequently sidestep celebration as well as warning, and probe instead the ontological question of what their objects are—what a medium or technology is—in essence, as well as the historical questions of how certain technologies or media got to be what they are. Yet perhaps one can further sidestep all of the above, not only celebration and warning but also the ontology and history of machines, toward a liberatory politics of formal expression; an attitude of friction when faced with new machines; an emancipatory cyberculture; an epistemological Luddism; a dismantling.

Marching against the Signs of Denial

As Houston Baker wrote of Shadrack, the founder of Suicide Day in Toni Morrison's *Sula*, political commitment may entail "marching against the very signs of one's denial,"[5] if one is to commit oneself to the critique of "global-technological"[6] modernity. Indeed, this risk of abnegation is what is undertaken in any public protest of consequence. It is what happens when a dissident group confronts the technologies of policing and occupation from a position of dissymmetrical power. To march against guns and tanks without guns and tanks of one's own is to engage in what is called "protest" rather than "war," and to move (as Huey Newton would say) against reactionary forces even at the risk of death. In short, under the rarest circumstances, a protest—like the Occupy Wall Street protests at Zuccotti Park or, to an even more pronounced degree, the Black Lives Matter actions in Ferguson, Baltimore, Baton Rouge, Oakland, and elsewhere—can involve the defiant survival of an act of revolutionary suicide. It is these, as acts of dismantling, that oppose the logics of American carnage.

Many of the most effective such acts have been photographed and distributed through social networks, like photos of the racial justice activists Shameeka Dream, Bree Newsome, and Ieshia Evans. The earliest of these photos, taken in 2015, shows Dream at the Baltimore protests following the police murder of Freddie Grey, walking the length of a line of riot police, holding a bundle of sage that is burning and smoking. The next photo, better known, is taken a month later and shows Newsome near the top of the flagpole on the lawn of the South Carolina statehouse, holding in her hand a Confederate flag that she has just removed from its halyard. The third and best known of these images, taken a year afterward, shows Evans in the middle of Airline Highway in Baton Rouge, protesting the police murder of Alton Sterling, Jr., by standing placidly in the path of police in thick body armor. In each of the three acts, an activist has been photographed in a position of extreme physical danger. Each risks death in the same kind of encounter that killed Grey, Sterling, and uncounted others. Yet each does so on purpose, facing the technology of state violence where Grey and Sterling had done so involuntarily; facing the technology of the journalist's camera where Grey and Sterling had been denied the chance.

If these actions are in risk-taking—in revolutionary suicide—then it is in part because they are so highly mediated. In calling them "mediated," I mean that they take their form at once through photography, in digital distribution of the resulting photographs, and by reference to history. As Teju Cole wrote in the *New York Times Magazine* about the photo of Evans: "She is unarmed and unafraid (the open space behind her emphasizes her singularity); they are militarized and unindividuated. The image told such an apparently clear story that when it hit social media, it went viral."[7] Each photo, thus distributed as something apparently clear, becomes exemplary of a kind of protest, coupling the human specificity of each activist with meanings so general as to be metaphorical. In this way, the activists' acts are mediated not only by the technologies of camera and social network but also by the politics of history and possibility. The image of Bree Newsome evokes the Iwo Jima Memorial in Arlington, Virginia. The image of Ieshia Evans evokes the Tank Man of Tiananmen Square, while suggesting as well the antiwar activist of 1967 who was photographed placing carnation stems in the barrels of National Guard rifles at the march on the Pentagon.

To these associations, Shameeka Dream's cleansing act of sage burning adds a much more recent association: it appears a kind of reversal of the 2011 pepper spraying of student protestors by officer John Pike, on the campus of the University of California, Davis. Through this thick web of associations, the images of Evans, Dream, and Newsome are linked both to a past imagination of radical citizenship and to the present reality of lives that could not be saved, and will not be saved. They signal also a possible deviation from that present, as the photographs continue (the Twitter profile of each activist foregrounds the picture in which she appears) to provide public certification of political credibility.

None of these activists died, and none has been killed in actions taken since. But their acts of civil disobedience were photographed in a time, now, when police racism regularly leads to a loss of life that is barely constrained and rarely punished. They could easily all have died, and they did what they did anyway, in the face of machines that are built only for death. This revolutionary suicide takes a form that, even though the events really happened, somehow seems even less literal than the fictional suicides of Russ, Morrison, and Pynchon that I explore in chapter 5. But their figuration on film, and their rhetorization as photographic figures, renders them just as transformative. If Houston Baker is correct that there is something Luddite about Morrison's Shadrack, then Newsome, Evans, and Dream repeat that Luddism in their defense of the conditions of subsistence. This is how critic R.H. Lossin describes it: "Now meaning an irrational dislike of all things new, Luddites historically are best known as machine breakers—smashing the stocking frames, gig mills, and automated looms that threatened not just their employment and wages, but a culture built around small shops, cottage industries, and craft guilds."[8] This is dismantling, not as machine hating but as a way to protect life against a large-scale regimentation and policing of security, labor, time, and community.

Dismantling is an opposition to mechanization and instrumentalization, and it signals a challenge to the epistemological limits of life under capitalism. It remains difficult but useful to imagine life with fewer machines, or to imagine confronting machines, like Ieshia Evans or Shameeka Dream staring down the barrel of a gun. Lossin warns: "An aversion to machine sabotage—in practice and in theory—is linked to an insufficient critique of technology and . . . the general weakness of such critiques is evidence of the symbolic and material importance of technology to a capitalist system of production."[9] Likewise, without the risk of death as one price of dismantling, political action is linked to an insufficient critique of the constraints on life. There are few practicable ways to think or act in the present as one who dismantles, who practices Luddism, or who risks death; but that does not mean that there are no practicable ways at all. Dismantling merely requires the refinement of a militant pacifism, a practice of self-suspension, technocritique, and protest, at the edge of the risk of death.

New Vultures

Consider, too, the 2016 protests against the Dakota Access Pipeline on the Standing Rock Indian Reservation, so far from Haudenosaunee lands, yet nevertheless legible in terms of what John Mohawk called liberation technology. From April 2016 until February 2017, indigenous and allied activists picketed the pipeline's construction site there as an effort to guard tribal land sites and waterways. On the side of the pipeline, its owners at Energy Transfer Partners, L.P., in alliance with governmental and business interests. On the other side, the so-called water-keepers whose appeals to environmental protections and religious liberties were coupled to organizing strategies that were owed to recent as well as much older struggles: from centuries-old anticolonial rebellions to the Indian Land Claims of the Long Seventies to the affinity-group activism of the Battle for Seattle, Occupy Wall Street, and Black Lives Matter.

The defense of the pipeline project from its protestors is explicitly a project that presumes a technological path toward freedom. As Richard Kauzlarich, an energy lobbyist and former Azerbaijan ambassador, wrote at the time: "Just as freedom is never more than one generation away from extinction, our energy independence must also be secured for the future generations. . . . Protests such as occurred in connection with the Dakota Access Pipeline must not determine public policy on this important question."[10] The connection is clear. National freedom is indissociable from energy independence. Energy independence, when not enabled by policy drives toward less devastating energy sources or less total energy expenditure, is indissociable from the fast transnational movement of raw materials. This fast transnational movement is enabled only by new technologies for transportation, like pipelines. These pipelines, in order to be built on contested lands, will need the institutional support of fully militarized (that is, fully technologized) state and national police enforcement. The cost of failure, in such a view, is high. Ayn Rand may as well have been talking about the Standing Rock protestors when she saw "the ecologists" as "the new vultures swarming to extinguish that fire" from a new machine.[11]

Contrarily, the defense of the land and waterways, as well as the resident community of working-class indigenous people, is explicitly a project of both technology and freedom, but it combines these terms in a very different way. Instead, to defend the land from the pipeline is to cultivate liberation technologies: exercised in opposition to those "technologies which generate centralization," on behalf of a world that is not "simply a commodity which can be exploited," such technologies are the procedures of community organizing, as well as those of historical recording that make it possible to synthesize contemporary organizing principles with those of the past. To be sure, digital networking was an important element of the organizing strategy in this case, as the hashtag #NoDAPL was applied to posts across social media, bringing together supporters and coordinating actions. Yet the

tools of social media were only ever at the service of more established organizing tools and historical tools. Moreover, when social media did become
the dominant form of organizing, they may have undermined the organizing on the ground. Ladonna Bravebull Allard, a leader of the Standing Rock
Sioux, explains how the technologies of state and energy interests have come
to collide with the technologies of communal self-determination, including
prayer and home:

> While we stand in prayer, we have assault rifles aimed at us, we are
> attacked by dogs, pushed from our sacred sites with pepper spray, shot
> with rubber bullets and bean bag rounds and Tasers, beaten with sticks,
> handcuffed and thrown in dog kennels. Our horses have been shot and
> killed. Our elders have been dragged out of ceremonies, our sacred
> bundles seized, our sacred eagle staff pulled from our hands. My daugh-
> ter was stripped naked in jail and left overnight for a traffic violation. An
> arsonist set the hills across from our camp afire, and for hours Morton
> County did nothing but prevent tribal authorities from responding.[12]

It is in the other actions taking place at Standing Rock, other than Twitter
or Instagram, other than the militarized police and the National Guard, and
other than the pipeline itself, that there is found an alternative way to think
about technology and freedom. The force of dismantling is here: where the
anticolonial liberation technologies (horses, prayer, bundles, staves, camps,
hills) are overcome by the technologies of governmentally enforced national
freedom (assault rifles, pepper spray, rubber bullets, dog kennels, bean bag
rounds, Tasers).

Thinking and Critique

To pick an opposed example, an unmanned aerial vehicle is a machine of
American carnage, a technology of tomorrow, and a tool to be dismantled.
Drones are the present instantiation of the technological capacity for mass
death at a distance; and when they are said to rain bombs on military and
civilian targets far away, drones are as a force of nature. When they are said to
exact violence by remote control, they are as a child's toy. While drones are
not in fact fully autonomous, they do provide relief to their human agents
as if they were autonomous. Theirs are not the actions of a god of rain, nor
those of a child at play, but instead of a machine that would run of itself.
As Jonathan Beller argues, the drone is not a rare or extreme case of contemporary technopolitical procedures, but is rather "a clear expression of
the general case in which visuality and the senses have been supervened by

computation and data-visualization to the extent that thinking and critique are short-circuited as subroutines subjugated to the programmed exigencies of machines."[13] Drones, like other machines, are the objects of a generalized condition of fetishization and disavowal. They do not remove the obligations to refuse or criticize, any more than any other machine might remove that obligation. Yet all the same, in the way that other machines go unrefused or uncriticized because they seem to be plaited into the fabric of the present, so too do drones (even when they are criticized) seem to be unrefusable.

Dismantling presses back at this process of disavowal, charting a thana-topography by which to deny drones along with the other technologies of mass death that have emerged since the conjunction of concentration camps to nuclear weapons. To undo the subvention of the senses by computation and data-visualization, in Beller's terms, is to admit that certain machines have soothed users into detaching themselves from responsibility for their own actions. It is to say loudly that those actions have consequences at plan-etary scale, and that indeed, it is those mediated and deadly actions that have given shape and scale to the planet. Under conditions of historical emergency (in the risk of death to human pilots; the risk of multiethnicity to national unity; the risk of a long war left unconcluded by a mushroom cloud) machines are granted emergency powers. But thinking and critique must not remain mere "subroutines" of this "programmed exigency," as if thinking and cri-tique were no match for technical procedures they deem ethically complex or geographically remote. These latter procedures can be dismantled, poetized, and deroutinized; repositioned on a longer timeline, a global map of the dead and dying.

Possibility of the Apocalypse

In 1957, in what might be the best-known touchstone for contemporary technological elites, Günther Anders wrote a series of dictates for a techno-logical age. Much reprinted throughout the Long Seventies, Anders's "Com-mandments in the Atomic Age" offers lessons that are specific to the develop-ment and proliferation of nuclear power and weaponry. It also offers a path to technological responsibility broadly speaking, in a world built on the near certainty of death and the slim but real chance of subsistence.

1. Your first thought upon awakening be: "Atom." You should not begin your day with the illusion that what surrounds you is a stable world.
2. The possibility of the Apocalypse is our work. But we know not what we are doing.

3. The futures which only yesterday had been considered unreachably far away, have now become neighboring regions of our present time.

4. Don't be a coward. Have the courage to be afraid.

5. The monopolistic claims for competency raised by individuals are unjustified because we all, as human beings, are equally incompetent.

6. You should not tolerate that the object, the effect of which surpasses all imagination, be classified by honest-sounding "keep smiling" labels.

7. Have and use only those things, the inherent maxims of which could become your own maxims, and thus the maxims of a general law.[14]

Anders's commandments compose an ethics by which to care for the self and others in ways that are strictly opposed to technological expertise. Anders does not command his readers, as the philosopher Babette Babich has remarked, so much as he offers "traditional spiritual exercises" and "rules for the direction of the soul, meditations of the Stoic kind."[15] These commandments on postnuclear survival are also ways to engage, on a daily basis, with the culture of invention and use that has filled every sector of modern life.

True, most historians know little about the operation of machines, even when they know much about their past applications. Equally true, most designers can tell you much about how to operate a machine but very little about its history or its interrelation with other machines. The object of Anders's depiction is an unstable world where most people are inexpert at most things, but where we all have to make decisions about a future apocalypse that lingers as the concurrent neighbor of the present. Be afraid, be very afraid, but adhere as well to this extension of Immanual Kant's categorical imperative: "Have and use only those things, the inherent maxims of which could become your own maxims, and thus the maxims of a general law."[16] To this starting point for any practical ethics, Anders adds that ethical maxims may inhere not only to human beings but also to their technological inventions. "Act as if the maxim of your action were to become by your will a universal law of nature," in Kant, thus becomes in Anders an imperative to have and use only the things whose intrinsic faculties might be had and used likewise by everyone else.

A decade and a half after Anders, feminist Charlotte Bunch writes her "Reform Tool Kit" in the first issue of Quest, the feminist journal she coedited. Rejecting reformism as mere accommodation, Bunch insists that institutions may yet be subject to reform, so as "to eliminate patriarchy and to create a more humane society."[17] To this end, without prescribing specific tools for use in all contexts, Bunch nevertheless advises feminists to examine all tools of reform according to five criteria. Phrased as questions, Bunch's criteria nevertheless recall Anders's earlier dicta:

> 1) Does this reform materially improve the lives of women, and if so, which women, and how many? 2) Does it build an individual woman's self-respect, strength, and confidence? 3) Does it give women a sense of power . . . [to] help build structures for further change? 4) Does it educate women politically, enhancing our ability to criticize and challenge the system in the future? 5) Does it weaken patriarchal control of society's institutions and help women gain power over them?[18]

Despite her different focus, Bunch can be seen gesturing to many imperatives that are seen in Anders. For him, technology is the very literal thing that will one day destroy the world; whereas for her, tools are the tactics of a movement with emancipatory aims. Still, the two are aligned in their shared logic of technological responsibility. In their common alignment, similar motivations reveal themselves: self-examination toward collective interest; collective orientation toward individual needs; modest ethical demands that face inordinate odds and therefore require large-scale political coordination; and an unwillingness to dispense with tools as such, even though most tools have so far been applied to the task of mass destruction.

Twenty-five years after Bunch and forty years after Anders—just as Anders had offered a modest extension of Kant, and Bunch suggests a specification of Anders's technological ethics for radical feminism—the philosopher Raphael Sassower also enumerates principles for the development and use of tools. Insisting on suggestions rather than commandments, but keeping Anders firmly in mind, Sassower poses a situational ethics of technoscientific knowledge. Whereas Anders had listed roughly seven dictates (depending on how one counts; with its overlaps and repetition, I count seven), and Bunch had asked five questions, Sassower's list is numbered from one to ten:

1. *Ask.* When signing up for a project/protocol, ask as many questions as possible.
2. *Answer.* Whatever answers you received, accept them provisionally and skeptically.
3. *Imagine.* Try to imagine the unimaginable.
4. *Think* . . . Push the logical boundaries to the breaking point.
5. *Quit.* Be ready to quit . . . because you may discover something offensive about the project/protocol.
6. *Dare to be skeptical.* Never think of projects/protocols as sacred.
7. *Divulge.* Never think of projects/protocols as secrets.

8. *Indulge.* Whenever possible, stray beyond the confines of your specialty.

9. *Explain.* Never think that what you do is beyond laypeople's comprehension.

10. *Remember.* Remember that whatever you are, you are still a member of the community.[19]

Sassower's list resembles the ones by Anders and Bunch, yet each of Sassower's suggestions is less a commandment or question than a plea. The force of each is in its clarity, and in the tragic and obvious fact that so few will follow it if it is ever read by anyone who is not already converted. On the whole, the list is not particularly philosophical. It is earnestly polemical, perfectly accessible, and not a little impatient, which after all is only appropriate to its moment and ours.

Decades have passed since the successive commandments of Anders, Bunch, and Sassower. If ever there were moments in which to write a book that pleads or makes demands, or asks antifoundational but collectivist questions, those were the moments. This is not such a moment. It is probably too late now anyway, when so few seem eager to hear what other people think they should do with their machines, when few seem willing either to admit their technical failings or to overcome them through technical training, and when few have enough knowledge and ethical consistency to be worth listening to. Yet even now, when rereading words written at the intersection of cyberculture and Luddism, one hears an injunction to readers that is as infectious as it is immodest in its insistence that we renew our obligations to one another even as technology changes around us. This infectious immodesty—a willingness by experts to speak clearly, or by nonexperts to speak righteously and poetically, about the culture of machines and expertise—may be the most valuable feature of Long Seventies literary politics. Falling squarely with Bunch, it hearkens back to Anders while foreshadowing Sassower. In this spirit, some propositions for dismantling, for the sake of survival and renewal—seven (like Anders) decades-old tactics that might be repeated or emulated, if nostalgia and cynicism do not bar the way:

1. Luddism: Letting go is neither difficult nor all that complicated, and we owe it to each other to break or relinquish any machine that kills our neighbors, regiments our communities, and partitions our movements toward justice. Some machines might be destroyed and some reprogrammed or repurposed, but others may simply be renounced, or left to falter, and then collectively ignored.

2. communion: Equally, it is not impossible to live together; to material-
ize relations between humans, between humans and nonhuman species, and between humans and machines. But communion, if it is ever
found, if it is even desirable, will likely resemble a conversation or a
riot, not a network, a spaceship, or a social media platform.

3. cyberculture: The tools of making are integrally tied to the tools by
which war is perpetrated and human life made fragile. If the world is
a poem, then it is a poem written from noncommunicative acts, both
of love and of terror, and not only from the ostensibly communicative acts from which planetary gestalts (like networks or markets) are
imagined.

4. distortion: The present is not only a runway for the launch of a future.
It is also a companion to a future that resides in this now; a bacchanal
in the graveyard of a past that is not gone; a workshop full of bones and
gravestones and broken bottles, arranged and rearranged, glued and
glittered into revolutionary renarrations of experience and purpose.

5. revolutionary suicide: There is no way to evade all complicity with the
principles that guide the design and deployment of such machines. Or
rather, there is no way to do so and survive. The horizon of such an
evasion is death. To survive is to accept some complicity, while to live a
good life is to minimize complicity while avoiding a meaningless death.

6. liberation technology: Most technological regimens have deserved
destruction, not because they are technological, but because of the
conditions of their designs and use. It is undesirable but not impossible to live without machines. Moreover, the anticipated end of cosmotechnics is not the end of everything. And almost miraculously,
some machines might not need to be destroyed at all.

7. thanatopography: To understand the relational character of the world is
to see how global social ethics must result from historical understandings
of the technologies of mass death more than from the technical mastery
of communication, computation, data storage, and travel.

What is called "the present" does not begin with the Trump administration
any more than it begins with the dawn of this no-longer-new twenty-first
century. The present does not precede the future; rather, the future (like its
past) distorts and neighbors the present.

Too often, enthusiasm for new technologies leads to a tacit acceptance
of the very imperatives that motivate technological invention. Even in spite
of ourselves, and even in our procedures of teaching and learning, we tend
regularly to endorse capitalism, militarism, cultural imperialism, and a naive

idea of individual freedom. We know very well that cell phones are built from toxic materials under exploitative conditions, but all the same we buy them. We know very well that data is stored on server farms with dire effects on local ecologies, but all the same we absorb and recite the fantasy of an immaterial "cloud" filled with immaterial information. We know very well that human difference is a primary condition of any ideal of togetherness, and not just its obstacle, but all the same we obey laws made by lawmakers who treasure telephones and the Internet for having made humans more culturally uniform. And yet technofetishism is not uniquely at fault for current conditions. Technophobes and primitivists make a parallel mistake when they blame new machines for social and political ills. As much as technophiles, technophobes put machines and technological invention at the center of their idea of contemporary life. But although the human present may indeed be inseparable from the invention of hardware, the programming of software, or the communication and storage of data, it is not reducible to these.

Totally New Problems

When Donald Trump announces the replacement of "American carnage" with the "technologies of tomorrow," he is not talking about real machines or about any machines that are not computational. He means only computers and he means them only in their least material and most ideological capacity. As Sarah Kember and Joanna Zylinska have shown, there is a "frequent conflation of 'new media' and 'new technology,'" such that the media, as analyzed by media theory, generally "become equated with the computer."[20] This slippage is reciprocal: just as new media are often conflated with new technology, so the word *technology* is generally taken as a synonym for new—that is digital—media. Even where it is acknowledged that social forms have changed through the introduction of political or industrial or military technologies, the signifiers of technology nevertheless increasingly stand in for the signifiers of computation and telecommunication. Because this is the case, there is a readerly barrier that frequently prevents the technology criticism of the past from being properly understood in the present. If a cyberculturist or a Luddite (or a thanatopographer, a liberation technologian, etc.) is talking about technology, but not about personal computers, mobile phones, drones, or the Internet, then what can she possibly contribute to the discourse of technology now?

And yet the computational discourse, when engaged, exerts a pull. The temptation is great, when enumerating technological violences in their historical or geospatial character, to say: *yes, but*. *Yes*, to destroy some machines or machinic habits is not the same as hating technology, even though it is the latter that most people would associate with dismantling. *Yes*, the promise of real coexistence was betrayed in favor of conference calls and video chats in a global village. *Yes*, actual uses of automation betrayed the potential of automation to lead instead toward politics and poetry. *Yes*, the distortion of normative histories is seen as destructive of a tradition-bound future rather than generative of new ways of living in the present. *Yes*, the technologies of freedom never turned out to be the ones leading toward planetary subsistence. *Yes*, political action is taken as good only when it carries no risk of death. *Yes*, we inherit an image of the world as a network of possibilities rather than a thanatopography of warnings. *But* maybe those technologies were used wrong and can yet be used right. To dismantle is to set aside this dithering of *yes, but* and to try instead the hard work of critique.

Exemplifying the failure to do this hard work, Andrew McAfee and Erik Brynjolfsson have lately tried to revivify the Triple Revolution as a model for prosperity in the new digital capitalism. Both professors of management at the Massachusetts Institute of Technology, McAfee and Brynjolfsson open their 2017 book with a chapter entitled "The Triple Revolution." In this reference to the letter from Alice Hilton, Linus Pauling, Robert Theobald, and others to Lyndon Johnson at the dawn of cyberculture, McAfee and Brynjolfsson repurpose earlier polemic toward ends that are much more optimistic and much less political. The three revolutions are no longer those of cybernation, weaponry, human rights; and the focus is no longer on the simultaneous risk and opportunity of the massive transitions in those fields. Instead, McAfee and Brynjolffson introduce a book that celebrates the convergence of developments in the three technologies listed in its title: *Machine, Platform, Crowd*. They write: "The successful companies of the second machine age will be those that bring together minds and machines, products and platforms, and the core and crowd very differently than most do today. Those that don't undertake this work, and that stick closely to today's technological and organizational status quo, will be the same choice as those that stuck with steam power."[21]

Is it a straight-faced parody or an accidental inversion that would twist ideas from that earlier coalition—anarchists, Marxists, and liberals joining up in defense of dignified work, peace, and racial justice—into a pop philosophy of business success? Platitudinous, McAfee and Brynjolfsson push early adoption and forward thinking as if these were themselves forward-thinking ideas. The book issues no caution about the new conditions, only an admission that "depending on how they are used, machines, platforms, and the crowd can have very different effects" such that "they can concentrate power and wealth or distribute decision making and prosperity."[22] Yet why should McAfee and Brynjolfsson not take the triple revolution as if it were their own formula, stripped of its prior associations with hard struggle? The previous triple revolution, after all, was a failed one. The new one might actually succeed. Total automation never did lead to a universal basic income. War and racism are, if anything, exacerbated by the new digital technologies. Why then should the idea of a triple revolution not be repurposed for computationalism and profit (either collective or individual) since no one was using it in the old way anymore?

No, there must be some life left in the triple revolution, in dismantling, in cyberculture and Luddism and at least some of the other terms that appear in this book. But if there is, it is likely of a modest sort, in their incitement to read or reread the literary politics of the Long Seventies, or in their invitation to rewrite the commandments of technological ethics for a distorted present. As autonomous terms, they provide no easy path to emancipation. They belong to the texts that generated them, which were polemical, unphilosophical, and very much of the struggles of their own time. They should be read in the present but not of the present. They provide no reason for the "optimistic vision for the future" that McAfee and Brynjolfsson derive from their own terms; and no reason to think with them that "the next few decades could and should be better than any other that humans have witnessed so far."[23] And yet the incitement to reread, like a confrontation with riot police with only a bundle of burning sage in hand, is not nothing. For a more fruitful return to the triple revolution, one need not look past 1978, in *Conversations in Maine: Exploring Our Nation's Future* by James Boggs, Grace Lee Boggs, Freddy Paine, and Lyman Paine. A handbook of sorts for radical thought and action, *Conversations in Maine* reformulates and updates the triple revolution without abandoning the radical spirit of the original. Both James and Grace Lee Boggs had participated in the first Conference on the Cybercultural Revolution, and James Boggs had been an original signatory to the Ad Hoc Committee's manifesto. Coming back to those ideas with the Paines, after fourteen years of antiracist labor organizing in Detroit, they leave two of the revolutions intact—"automation and cybernation [which] pose totally new problems of work" and "military technology [which] poses the end of the nation-state and war"—while building into the human rights revolution some insights from feminist struggle of the intervening years: "the third problem area is the technology of reproduction, of maternity and of child care . . . [so] that unless women find a new role for themselves, they will stagnate."[24]

The Boggses and Paines, in their return to the triple revolution, argue that the right to abortion and the improvements to contraceptive and pediatric medicine now necessitate a total social reorganization. Although less radical (or else simply less optimistic) in their vision than Shulamith Firestone, they nevertheless echo Firestone's insistence that gender roles will change partly due to changed practices of automation and peace: "The technological revolution has created critical questions in three spheres . . . and they are interrelated. They pose the possibility of our moving again to the evolution of humankind—as before the rise of the state; the possibility of both men and women contributing equally as they have not been able to do for the last five thousand years."[25] It is strange indeed to ask a study in contemporary management to measure up to a critique of the state, especially in an activist handbook like *Conversations in Maine*. Yet it is McAfee and Brynjolfsson, in appropriating the triple revolution for themselves, who make the comparison visible. By cribbing the phrase while emptying it of emancipatory principles, McAfee and Brynjolfsson can only promote their machines, the digital capitalism to which they contribute. By contrast, the Boggses and Paines embrace the terms of the prior moment, the earliest years of the Long Seventies, while attempting to equip their struggle toward new challenges in the dark times to come. When the Paines and Boggses incorporate feminist language into their work as labor and racial justice organizers, they thereby enter a coalitional practice like the one announced by Audre Lorde at around the same time. To print their dialogue in a book, a conversation, a discursive experiment in activist language, is to participate in a literary politics. The words of *Conversations in Maine* should not, as the words of Lorde and Firestone should not, be read as if they were scripture. Exegesis is a terrible answer, the worst kind of descriptive reading. But some answer may lie in these words nevertheless. They adapt the ideas that precede them, and they survive as provocations to adapt other prior ideas, and other extant machines, even at the risk of destroying them.

It can therefore be acknowledged with the Boggses and Paines that "unbridled technology is catastrophic to man's future"; and that at once "it is equally obvious that what man has learned about the uses of technology should not, cannot ever be lost."[26] Yet acknowledging the need to be wary is not the same as using caution to found a politics of dismantling in the present. In the twenty-first century, a time and place far from the Maine conversations of the Boggses and Paines, there must be a willingness to lose the machines. To decide in advance not to lose them is only the least objectionable version of the ambivalent formula: *yes, but. Yes*, technology should not be "unbridled"; and *yes*, the best of us want something other than a "technological revolution"; *but* the value of certain machines "cannot ever be lost." It was a powerful ambivalence in its moment, because it called out unbridled exploitation but held firm to technological possibility. But at this late date, when tech firms and state agents hold all the codes, something still more urgent is required: to sweep past noncommittal logics, to risk relinquishing any use of any machine (even the mechanized self), and to move in groups toward more concrete practices of technological responsibility through making and breaking. That is, through dismantling.

Acknowledgments

This book owes its completion to two people: Anita Starosta, dearest comrade in ideas and everything else; and Brian Lennon, model of friendship and good sense. Susan Webster Tierney's teaching persists everywhere here, in a way that is different in kind but not degree from the teachings of Teresa de Lauretis, Rey Chow, Michael Silverman, Carla Freccero, David Marriott, Barry Maxwell, and (most indelibly) Ellen Rooney. No editor is a truer ally than Mahinder Kingra, and none is so conscientious as Sheila Marie Flaherty-Jones. It was a unique pleasure to work again at the side of Ange Romeo-Hall. At Penn State, I am grateful for the support of my department head Mark Morrisson, as well as the dean of the College of the Liberal Arts, the Humanities Institute, and the English Department staff. Tina Chen convened a panel of colleagues—Bishnupriya Ghosh, Kevin Bell, Claire Colebrook, Brian Lennon, and Tina herself—whose input at the manuscript stage was decisive as well as impossibly generous.

For key conversations, I thank Alexis Shotwell, Leah Paulos, Rich Purcell, Hong-An Tran, Carlos Amador, Nick Mitchell, Beth Capper, Sarah Osment, Anthony Reed, and my cadre on Twitter. Locally, it was Ebony Coletu, Anne McCarthy, John Schneider, Susan Squier, Max Larson, Laura Jones, Sam Tenorio, Jon Abel, Brandon Pettit, Jeff Nealon, Janet Lyon, Chris Castiglia, Jonathan Fedors, John Marsh, Leisha Jones, Carla Mulford, Daniel Purdy, Chris Reed, Julia Kasdorf, Ben Schreier, Aldon Nielsen, Magalí Armillas-Tiseyra, Paul Kellerman, Stuart Selber, Nergis Ertürk, Rachel Wiley, Sarah Townsend, Sam Frederick, Casey Wiley, Anna Ziajka-Stanton, Michael Bérubé, Rich Doyle, Chris Stanton, Alex Fattal, Garrett Sullivan, Hester Blum, Bruno Jean-François, Julie Kleinman, or (much more often than not) Claire Bourne who walked with me through an idea or toward a sense of common endeavor. I thank Jim Tierney and Liz Strout for their imperishable support; and my siblings—Katie Tierney, Dan Tierney, Josie Tierney-Fife, Adam Tierney-Eliot, Hanne Tierney, Pete Tierney-Fife, Allison Nelson-Eliot, Mikey Litwack, Cheryl Beredo, and Descha Daemgen—for theirs.

An early version of material in chapters 3 and 4 appeared as "Cyberculture in the Large World House" in *Configurations* 26, no. 2 (2018). Ian Miller, a genuine giant of book illustration and a delight of a correspondent, made the image that appears on the cover. Three online collections—Independent Voices, American Indian Histories and Cultures, and the Internet Archive—provide access to rare and indispensable texts. Joel Simundich saved my life.

This book owes much to an industrywide division of labor that awards the expectation of research to a small minority of scholars, while denying that privilege, along with security and pay equity, to an enormous and rapidly growing majority. To the incalculable extent that this division of labor benefits me in this time of writing, I can only acknowledge my debt with humility and solidarity.

In all other respects, this book and I are dedicated to students.

Notes

Introduction: For the Sake of Survival

1. Alice Mary Hilton, *Logic, Computing Machines, and Automation* (Washington, DC: Spartan Books, 1963), 398.

2. Siegfried Zielinski, . . . *After the Media* (Minneapolis: University of Minnesota Press, 2013), 173.

3. Zielinski, . . . *After the Media*, 173.

4. Zielinski, . . . *After the Media*, 14.

5. Walter Benjamin, *Origin of the German Trauerspiel*, trans. Howard Eiland (Cambridge, MA: Harvard University Press, 2019), 10–11.

6. Audre Lorde, *Sister Outsider: Essays and Speeches* (Trumansburg, NY: Crossing Press, 1984), 110.

7. Hilton, *Logic, Computing Machines, and Automations*, 373–374.

8. Sara Ahmed, *Living a Feminist Life* (Durham, NC: Duke University Press, 2017), 2.

9. Mike Davis et al., quoted in Louis Proyect, "Robert Brenner, Vivek Chibber, and the 'Organization Question,'" *Louis Proyect: The Unrepentant Marxist* (blog), June 25, 2018, https://louisproyect.org/2018/06/25/robert-brenner-vivek-chibber-and-the-organization-question/.

10. Lane Windham, *Knocking on Labor's Door: Union Organizing in the 1970s and the Roots of a New Economic Divide* (Chapel Hill: University of North Carolina Press, 2017), 5.

11. Cal Winslow, "Overview: The Rebellion from Below, 1965–1981," in *Rebel Rank and File: Labor Militancy and Revolt from Below during the Long 1970s*, ed. Aaron Brenner, Robert Brenner, and Cal Winslow (New York: Verso, 2010), 3–4.

12. Howard Brick and Christopher Phelps, *Radicals in America: The U.S. Left since the Second World War* (Cambridge: Cambridge University Press, 2015), 1:216–217.

13. H. Bruce Franklin, *Robert A. Heinlein: America as Science Fiction* (Oxford: Oxford University Press, 1980), 199.

14. Norbert Wiener, "A Scientist Rebels," *The Atlantic*, January 1947, 46.

15. Norbert Wiener, *Cybernetics, or Control and Communication in the Animal and the Machine* (Cambridge, MA: MIT Press, 1948), 28–29.

16. Marshall McLuhan, introduction to *Explorations in Communication*, ed. Marshall McLuhan and Edmund Carpenter (Boston: Beacon Press, 1960), xi.

17. Marshall McLuhan, "Interview with Gerald Emanuel Stearn," in *McLuhan, Hot and Cool*, ed. Gerald Emanuel Stearn (New York: Dial Press, 1967), 279.

18. McLuhan, "Interview," 272.

19. Kenneth Burke, *Language as Symbolic Action* (Berkeley: University of California Press, 1966), 413.

20. Raymond Williams, *The Country and the City* (New York: Oxford University Press, 1975), 295–296.

21. R. Buckminster Fuller, "The Comprehensive Man," *Northwest Review* 2, no. 2 (1959): 24.

22. Fuller, "Comprehensive Man," 24.

23. Fuller, "Comprehensive Man," 29.

24. Stewart Brand, *Whole Earth Catalog* (Menlo Park: Portola Institute, 1968), 2.

25. Albert Szent-Györgyi, *What Next?!* (New York: Philosophical Library, 1971), 60.

26. In Alice Mary Hilton, ed., *The Evolving Society: The Proceedings of the First Annual Conference on the Cybercultural Revolution* (New York: Institute for Cybercultural Research, 1966), 234.

27. In Dorothy Bates, "The Machine I Hate the Most," *Avant Garde*, 1971, 23.

28. Karl Marx, *Capital: A Critique of Political Economy*, trans. Ben Fowkes (New York: Penguin Books, 1990), 1:554–555.

29. Eric J. Hobsbawm, "The Machine Breakers," *Past & Present* 1 (1952): 58.

30. Hobsbawm, "Machine Breakers," 66.

31. E.P. Thompson, *The Making of the English Working Class* (New York: Vintage Books, 1963), 601.

32. Hobsbawm, "Machine Breakers," 67.

33. John Zerzan and Paula Zerzan, "Who Killed Ned Ludd?," *Fifth Estate*, April 1976, 15.

34. Zerzan and Zerzan, "Who Killed Ned Ludd?," 7.

35. In Bates, "Machine I Hate the Most," 23.

36. Kenneth Burke, *Late Poems 1968–1993: Attitudinizings Verse-Wise, while Fending for One's Selph, and in a Style Somewhat Artificially Colloquial*, ed. Julie Whitaker and David Blakesley (Columbia: University of South Carolina Press, 2005), 100.

37. Jipson John, Jitheesh P.M., and David Harvey, "'The Neoliberal Project Is Alive but Has Lost Its Legitimacy': David Harvey," *The Wire*, February 9, 2019, https://thewire.in/economy/david-harvey-marxist-scholar-neo-liberalism.

38. Kristie Dotson, "How Is This Paper Philosophy?" *Comparative Philosophy* 3, no. 1 (2012): 25–26.

39. Martin Heidegger, *The Question concerning Technology and Other Essays*, trans. William Lovitt (New York: Harper and Row, 1977), 13.

40. Jean-François Lyotard, *The Postmodern Condition: A Report on Knowledge*, trans. Geoff Bennington and Brian Massumi (Minneapolis: University of Minnesota Press, 1984), 44.

41. Lyotard, *Postmodern Condition*, 45.

42. Lyotard, *Postmodern Condition*, 47.

43. Lyotard, *Postmodern Condition*, 44.

44. Theodor W. Adorno, "Education after Auschwitz," in *Critical Models: Interventions and Catchwords*, trans. Henry W. Pickford (New York: Columbia University Press, 1998), 200.

45. Adorno, "Education after Auschwitz," 201.

46. Theodor W. Adorno, "Late Capitalism or Industrial Society? The Fundamental Question of the Present Structure of Society," in *Can One Live after Auschwitz? A Philosophical Reader* (Palo Alto: Stanford University Press, 2003), 118.

47. Karl Marx, "Theories of Surplus Value," in *The Collected Works of Karl Marx and Friedrich Engels* (London: Lawrence and Wishart, 2010), 33:289.

48. Karl Marx, *Grundrisse: Foundations of the Critique of Political Economy*, trans. Martin Nicolaus (New York: Penguin, 1973), 699.

49. Kenneth Burke, "Why Satire, with a Plan for Writing One," *Michigan Quarterly Review* 13, no. 4 (1974): 331.

50. Moishe Postone, "Necessity, Labor, and Time: A Reinterpretation of the Marxian Critique of Capitalism," *Social Research* 45, no. 4 (1978): 778.

51. Postone, "Necessity, Labor, and Time," 778.

52. Postone, "Necessity, Labor, and Time," 778. Postone's emphasis.

53. David Fenton and Gil Scott-Heron, "'The First Minute of a New Day': Music and Politics with Gil Scott-Heron," *Ann Arbor Sun*, March 14, 1975, 19.

54. Stephen Paul Miller, *The Seventies Now: Culture as Surveillance* (Durham, NC: Duke University Press, 1999), 11–12.

55. Kristin Ross, *May '68 and Its Afterlives* (Chicago: University of Chicago Press, 2002), 8.

56. Ross, *May '68 and Its Afterlives*, 24.

57. Elizabeth A. Freeman, *Time Binds: Queer Temporalities, Queer Histories* (Durham, NC: Duke University Press, 2010), xiii.

58. Freeman, *Time Binds*, xiv.

1. Luddism

1. W.S. Merwin, *The Miner's Pale Children: Prose* (New York: Atheneum, 1969), 131.

2. Dominic Pettman, "The Species without Qualities: Critical Media Theory and the Posthumanities," *b2o: the online community of the boundary 2 editorial collective*, April 23, 2019, http://www.boundary2.org/2019/04/the-species-without-qualities-critical-media-theory-and-the-posthumanities/.

3. In this and other observations on the technological character of Lorde's "master's tools" metaphor, I owe much to a conversation with student members of the Reading for the Future Collective at Penn State, particularly Miriam Gonzales, Erica Stevens, and Chris Chidi.

4. Audre Lorde, *Sister Outsider: Essays and Speeches* (Trumansburg, NY: Crossing Press, 1984), 112.

5. Lorde, *Sister Outsider*, 110.

6. Lorde, *Sister Outsider*, 111.

7. For a genealogy of this association, posed as a history of metaphors rather than of concepts, see "Organic and Mechanical Background Metaphorics," chapter 6 of Hans Blumenberg's 1960 book *Paradigms for a Metaphorology*.

8. Lorde, *Sister Outsider*, 110–111.

9. Charlotte Bunch, "The Reform Tool Kit," *Quest: A Feminist Quarterly* 1, no. 1 (1974): 40.

10. Bunch, "Reform Tool Kit," 43.

11. Bunch, "Reform Tool Kit," 40.

12. Lorde, *Sister Outsider*, 37–38.

13. Roderick A. Ferguson, "Of Sensual Matters: On Audre Lorde's 'Poetry Is Not a Luxury' and 'Uses of the Erotic,'" *Women's Studies Quarterly* 40, no. 3/4 (2012): 300.

14. Lorde, *Sister Outsider*, 38.

15. Ferguson, "Of Sensual Matters," 296.

16. Audre Lorde, *From a Land Where Other People Live* (Detroit: Broadside Press, 1973), 7.

17. Lorde, *From a Land Where Other People Live*, 7.

18. Lorde, *From a Land Where Other People Live*, 7.

19. John Ashbery, *Selected Poems* (New York: Viking Penguin, 1985), 18.

20. Elizabeth Povinelli, *Geontologies: A Requiem to Late Liberalism* (Durham, NC: Duke University Press, 2015), 28.

21. K. Wayne Yang provides an alternative reading of these lines—"Listen: you will need to remember this when you are accused of destruction. Attach a pacemaker to the heart of those machines you hate; make it pump for your decolonizing enterprise; let it tick its own countdown. Ask how, and how otherwise, of the colonizing machines. Even when they are dangerous"—writing under a pseudonym, la paperson, in *A Third University Is Possible* (Minneapolis: University of Minnesota Press, 2017), 24.

22. Lorde, *From a Land Where Other People Live*, 8.

23. For a different take on the continuing utility of Lorde's metaphor in the present, see the excellent 2006 collection edited by Lewis Gordon and Jane Anna Gordon, *Not Only the Master's Tools: African American Studies in Theory and Practice*.

24. Langdon Winner, *Autonomous Technology: Technics-out-of-Control as a Theme in Political Thought* (Cambridge, MA: MIT Press, 1977), 251.

25. Winner, *Autonomous Technology*, 190.

26. Winner, *Autonomous Technology*, 263–264.

27. Winner, *Autonomous Technology*, 330. Winner's emphasis.

28. Winner, *Autonomous Technology*, 207.

29. Winner, *Autonomous Technology*, 331.

30. In Kevin Binfield, ed., *Writings of the Luddites* (Baltimore: Johns Hopkins University Press, 2004), 99.

31. Dorothy Bates, "The Machine I Hate the Most," *Avant Garde* (1971): 22.

32. Bates, "Machine I Hate the Most," 23.

33. Bates, "Machine I Hate the Most," 23.

34. Bates, "Machine I Hate the Most," 23.

35. Édouard Glissant, *Caribbean Discourse: Selected Essays*, trans. Michael Dash (Charlottesville: University of Virginia Press, 1989), 99–100.

36. Édouard Glissant, *Poetic Intention*, trans. Nathalie Stephens (Callicoon, NY: Nightboat Books, 2010), 22.

37. Glissant, *Poetic Intention*, 159.

38. William V. Spanos, "Breaking the Circle: Hermeneutics as Dis-closure," *Boundary 2* 5, no. 2 (1977): 446.

39. Paul A. Bové, *Destructive Poetics: Heidegger and Modern American Poetry* (New York: Columbia University Press, 1980).

40. Winner, *Autonomous Technology*, 326.

41. Winner, *Autonomous Technology*, 331. Winner adds here, however: "Perhaps Paul Goodman was [an epistemological Luddite] on occasion."

42. 118 Cong. Rec., no. 4 (1972), 4803.

43. 118 Cong. Rec., 4806.

44. 118 Cong. Rec., 4806.

45. Robert Scholes, *Fabulation and Metafiction* (Urbana: University of Illinois Press, 1979), 131.

46. Merwin, *Miner's Pale Children*, 29.

47. Merwin, *Miner's Pale Children*, 131.

48. Merwin, *Miner's Pale Children*, 56.

49. Merwin, *Miner's Pale Children*, 58.

2. Communion

1. Lewis Mumford, *Technics and Civilization* (Chicago: University of Chicago Press, 2010), 110. Mumford delineated four stages in the history technology, whereas Szent-Györgyi importantly differentiates cosmotechnics only from paleotechnics and neotechnics. For Mumford, paleotechnics was preceded by eotechnics—"a water-and-wood complex"—and neotechnics would be followed by biotechnics: a phase of "completer integration of the machine with human needs and desires" that lay "already visible over the edge of the horizon" (353). There is no reason to think eotechnics at all incompatible with Szent-Györgyi's periodization, since it simply precedes it. But his preference for cosmotechnics over biotechnics suggests a difference of focus from that of Mumford. Whereas Mumford described the technological transformations of human bodies and practices, Szent-Györgyi described the technological transformations of commonly shared assumptions about the unity and picturability of the world.

2. This notion of cosmotechnics, while it is not cited there, shares much with Yuk Hui's much later use of this word in the philosophy of technology, as it is developed in "On Cosmotechnics: For a Renewed Relation between Technology and Nature in the Anthropocene," *Techné: Research in Philosophy and Technology* 21, no. 2/3 (2017): 319–341.

3. Albert Szent-Györgyi, *What Next?!* (New York: Philosophical Library, 1971), 26. Cosmotechnics might also be called *communicative globalism* if one were to adapt a term developed by Jodi Dean, *communicative capitalism*. The latter phrase signals a stage in twenty-first-century capitalism that is held in place partly through a peculiar, illusory form of democratic participation in which every form of expression is definable as a message that has been sent and received. Communicative globalism, or cosmotechnics, would then be the worldview that emerges as consequence. See Dean's *Democracy and Other Neoliberal Fantasies: Communicative Capitalism and Left Politics* (2009).

4. Szent-Györgyi, *What Next?!*, 48.

5. Alvin M. Weinberg, "Can Technology Replace Social Engineering?" *Bulletin of the Atomic Scientists* 22 (1966): 5.

6. Szent-Györgyi, *What Next?!*, 42.

7. Szent-Györgyi, *What Next?!*, 43.

8. R. Buckminster Fuller, "Man's Total Communication System," *Evergreen Review* 14, no. 83 (1970): 39.

9. Fuller, "Man's Total Communication System," 64.

10. Fuller, "Man's Total Communication System," 61.

11. Fuller, "Man's Total Communication System," 61.

12. R. Buckminster Fuller, *Operating Manual for Spaceship Earth* (Carbondale: Southern Illinois University Press, 1968), 19.

13. Adlai E. Stevenson, *The Papers of Adlai E. Stevenson*, vol. 8, ed. Walter Johnson (Boston: Little, Brown, 1979), 828. Stevenson's emphasis.

14. Stevenson, *Papers of Adlai E. Stevenson*, 8:828. Stevenson's emphasis.

15. Stevenson, *Papers of Adlai E. Stevenson*, 8:828. Stevenson's emphasis.

16. Garrett Hardin, "Living on a Lifeboat," *BioScience* 24, no.10 (1974): 561.

17. Hardin, "Living on a Lifeboat," 561.

18. In Stevenson, *Papers of Adlai E. Stevenson*, 8:805.

19. Stevenson, *Papers of Adlai E. Stevenson*, 8:810.

20. Stevenson, *Papers of Adlai E. Stevenson*, 8:812.

21. Barbara Ward, *Spaceship Earth* (New York: Columbia University Press, 1966), vii.

22. Ward, *Spaceship Earth*, 130.

23. Alexander R. Galloway, with David M. Berry, "A Network Is a Network Is a Network: Reflections on the Computational and the Societies of Control," *Theory Culture & Society* 33, no. 4 (2016): 160.

24. Galloway, with Berry, "Network Is a Network," 160.

25. Ursula K. Le Guin, "The Child and the Shadow," *Quarterly Journal of the Library of Congress* 32, no. 2 (1975): 141–143.

26. Mike Hodel, "A Talk with Philip K. Dick," interview with Philip K. Dick, North Hollywood, KPFK-FM Radio Network, June 26, 1976.

27. Joanna Russ, "Books," *Fantasy & Science Fiction*, March 1975, 43.

28. Fredric Jameson, "World-Reduction in Le Guin: The Emergence of Utopian Narrative," *Science Fiction Studies* 2, no. 3 (1975): 223.

29. Jameson, "World-Reduction in Le Guin," 228.

30. Ursula K. Le Guin, "The Algol Interview: Ursula K. Le Guin," *Algol* 12, no. 2 (1975): 9.

31. Le Guin, "Algol Interview," 9–10.

32. Ursula K. Le Guin, *The Dispossessed: An Ambiguous Utopia* (New York: Perennial Classics, 1974), 278.

33. Le Guin, *Dispossessed*, 279.

34. Le Guin, *Dispossessed*, 280.

35. Le Guin, *Dispossessed*, 221.

36. Le Guin, *Dispossessed*, 280.

37. M.M. Bakhtin, *The Dialogic Imagination: Four Essays*, ed. Michael Holquist, trans. Caryl Emerson and Michael Holquist (Austin: University of Texas Press, 1981), 157.

38. Le Guin, *Dispossessed*, 281.

39. Le Guin, *Dispossessed*, 344.

40. Samuel R. Delany, introduction to *The Cosmic Rape and "To Marry Medusa"* by Theodore Sturgeon (Boston: Gregg Press, 1977), xxiii.

41. In Samuel R. Delany, *Triton: An Ambiguous Heterotopia* (New York: Bantam Books, 1976), 343.

42. Delany, *Triton*, 14.

43. Delany, *Triton*, 22–23.

44. Delany, *Triton*, 113.

45. Delany, *Triton*, 227.

46. Delany, *Triton*, 228.

47. Delany, *Triton*, 228.

48. Samuel R. Delany, "The *Algol* Interview: Samuel R. Delany," *Algol* 13 (1976): 18.

49. Lauren Berlant, *The Female Complaint: The Unfinished Business of Sentimentality in American Culture* (Durham, NC: Duke University Press, 2008), 203–204.

50. Michel Foucault, *The Archaeology of Knowledge and the Discourse on Language*, trans. A.M. Sheridan Smith (New York: Pantheon Books, 1972), 225.

51. Foucault, *Archaeology of Knowledge*, 225.

52. William D. Seidensticker, "Language as Communication: A Criticism," *Southwestern Journal of Philosophy* 2, no. 3 (1971): 31.

53. Seidensticker, "Language as Communication," 31.

54. "Constitution of UNESCO," November 16, 1945, UNESCO, http://portal.unesco.org/en/ev.php-URL_ID=15244&URL_DO=DO_TOPIC&URL_SECTION=201.html.

55. Herbert I. Schiller, "Freedom from the 'Free Flow,'" *Journal of Communication* 24, no.1 (1974): 116.

56. Armand Mattelart, "Modern Communication Technologies and New Facets of Cultural Imperialism," *Instant Research on Peace and Violence* 3, no. 1 (1973): 9.

57. Mattelart, "Modern Communication Technologies," 23.

58. Armand Mattelart, *Transnationals and the Third World* (South Hadley, MA: Bergin & Garvey, 1983), 132.

59. Archibald MacLeish, "A Reflection: Riders on the Earth Together, Brothers in Eternal Cold," *New York Times,* December 25, 1968, A1.

60. Jean-Pierre Dupuy, "Myths of the Informational Society," in *The Myths of Information: Technology and Postindustrial Culture*, ed. Kathleen Woodward (Madison, WI: Coda Press, 1980), 5.

61. Dupuy, "Myths of the Informational Society," 5.

3. Cyberculture

1. Joseph Tirella, *Tomorrow-Land: The 1964–65 World's Fair and the Transformation of America* (Guilford, CT: Lyons Press, 2014), 212.

2. Talk of the Town, *The New Yorker*, July 4, 1964, 25.

3. Talk of the Town, 22.

4. Talk of the Town, 22.

5. Alice Mary Hilton, "Cybernetics and the Future," interview with Studs Terkel, Chicago, WFMT Radio Network, July 12, 1965.

6. Hannah Arendt, *The Human Condition*, 2nd ed. (Chicago: University of Chicago Press, 1998), 9.

7. Hannah Arendt, *The Portable Hannah Arendt*, ed. Peter Baehr (New York: Penguin, 2000), 168.

8. Arendt, *Portable Hannah Arendt*, 168.

9. Of note, the Ad Hoc Committee's image of the future foreshadows other utopian images, from science-fictional elaborations to accelerationism, as this latter

is theorized on the Right by Nick Land and on the Left by Nick Srnicek and Alex Williams, and as these elaborations trickle into the industrious pseudoscience of Elon Musk and other technologists. Like cyberculture critique, accelerationism suggests that advantages may result from a commitment to, rather than a rejection of, forms of computation that displace human labor. However, where the accelerationist sees a capitalism that will devour itself and its societies, the triple revolutionist or cyberculturist sees a capitalism that will suddenly, with the advent of abundance, cease to make sense. For their part, Srnicek and Williams have dismissed the Triple Revolution in a footnote, including the manifesto in a list of several documents that have, since the 1960s, expressed a "fear of automation taking jobs."

10. In Alice Mary Hilton, ed., *The Evolving Society: The Proceedings of the First Annual Conference on the Cybercultural Revolution* (New York: Institute for Cybercultural Research, 1966), 234.

11. Alice Mary Hilton, "Hilton's Law," *Improving College and University Teaching* 16, no. 2 (1968): 149.

12. James Perrin Warren, "'Catching the Sign': Catalogue Rhetoric in 'The Sleepers,'" *Walt Whitman Quarterly Review* 5 (1987): 19.

13. Hilton, "Hilton's Law," 149.

14. Alice Mary Hilton, "Individual Responsibility in a Cybercultural Society," *Quest* 5 (1965): 37.

15. Hilton, "Cybernetics and the Future."

16. Norbert Wiener was himself not particularly fond of the word *cyberculture*, and pled with Hilton in a letter one year before his death: "Why can't you call it 'Culture in Cybernetics'? Or 'Culture in Communication'? You will pardon me for my criticizing, but these portmanteau words rub me the wrong way and they sound to me like a streetcar making a turn on rusty rails." For her part, Hilton was quite aware that the word was grating to the ear, chalking this effect up to its casual conjunction of a Greek word with a Latin word. The word's disjunctive sound, she thought, was an echo of its disjunctive social effects. (Wiener quoted in Thomas Rid, *Rise of the Machines: A Cybernetic History* [New York: W.W. Norton, 2017], 103.)

17. Hilton, "Hilton's Law," 149.

18. In seeking an explanation for why contemporary definitions of *cyberculture* differ so widely from those of its earliest appearance, it is worth consulting another critic at the millennium, David Bell, for whom "the birthplace and birthdate of the term 'cyberculture' is . . . obscure and uncertain . . . although it was being used quite widely in academia by the mid-1990s." This is a startling claim, given the ready availability of dictionaries that list Hilton's coinage. But perhaps it is not so startling if it is motivated by a desire to shore up existing usage. An exclusively digital take on the term might be found applicable to pernicious old problems of class and race, gender and labor, but it could hardly be seen as having emerged from the critique of those very problems.

19. Fred Turner, *From Counterculture to Cyberculture: Stewart Brand, the Whole Earth Network, and the Rise of Digital Utopianism* (Chicago: University of Chicago Press, 2006), 33.

20. Turner, *From Counterculture to Cyberculture*, 38.

21. Alice Mary Hilton, *Logic, Computing Machines, and Automation* (Washington, DC: Spartan Books, 1963), xvi.

22. Hilton, *Logic, Computing Machines, and Automation*, 27.

23. Hilton, *Logic, Computing Machines, and Automation*, 378.

24. Ad Hoc Committee, "The Triple Revolution: Cybernation—Weaponry—Human Rights," in *Seeds of Liberation*, ed. Paul Goodman (New York: George Braziller, 1964), 412–413.

25. Heather Hicks, *The Culture of Soft Work: Labor, Gender, and Race in Postmodern American Narrative* (New York: Palgrave MacMillan, 2009), 105.

26. For a rich discussion of racial justice as a motivating factor in these early discussions of cyberculture, especially as regards the contributions of the Boggses, see Michael Litwack's "Racial Technics: Media and Machines in the Long Civil Rights Era" (PhD diss., Brown University, 2016), especially the chapter entitled "Working (Like) the Machine: Life, Labor, and Cybernation in the Motor City."

27. On technological solutionism, see Evgeny Morozov, *To Save Everything, Click Here: The Folly of Technological Solutionism*; and, from a more theoretical perspective, David Golumbia, *The Cultural Logic of Computation*.

28. Hilton, "Cybernetics and the Future."

29. Ad Hoc Committee, "Triple Revolution," 412–413.

30. Hilton, *Logic, Computing Machines, and Automation*, 378.

31. Don Ihde, *Technics and Praxis* (Dordrecht: D. Reidel, 1979), 109.

32. N. Bruce Hannay and Robert E. McGinn, "The Anatomy of Modern Technology: Prolegomenon to an Improved Public Policy for the Social Management of Technology," *Daedalus* 109, no. 1 (1980): 27.

33. Andrew Feenberg, "The Political Economy of Social Space," in *The Myths of Information: Technology and Postindustrial Culture*, ed. Kathleen Woodward (Madison, WI: Coda Press, 1980), 113.

34. Langdon Winner, *Autonomous Technology: Technics-out-of-Control as a Theme in Political Thought* (Cambridge, MA: MIT Press, 1977), 100.

35. Albert Borgmann, "The Explanation of Technology," *Research in Philosophy and Technology* 1 (1978): 99.

36. Mario Bunge, "The Five Buds of Technophilosophy," *Technology in Society* 1 (1979): 68.

37. Winner, *Autonomous Technology*, 3.

38. Murray Bookchin, *Post-Scarcity Anarchism*, 2nd ed. (Montreal: Black Rose Books, 1986), 122.

39. Bookchin, *Post-Scarcity Anarchism*, 152.

40. Herbert Marcuse, "The End of Utopia," *Ramparts*, April 1970, 32.

41. Theodore Roszak, *The Making of a Counter Culture: Reflections on the Technocratic Society and Its Youthful Opposition* (Garden City, NY: Anchor Books, 1969), 280.

42. Paul Goodman, "The Trouble with Today's Technology: A Social Critic's View," *Innovation*, June 1969, 39.

43. Goodman, "Trouble with Today's Technology," 39.

44. George Jackson, *Soledad Brother: The Prison Letters of George Jackson*, ed. Jonathan Jackson, Jr. (Chicago: Lawrence Hill Books, 1994), 244.

45. Jackson, *Soledad Brother*, 262.

46. Derrick de Kerckhove, *The Skin of Culture: Investigating the New Electronic Reality* (Toronto: Somerville House, 1995), 137.

47. Pierre Lévy, *Cyberculture*, trans. Robert Bononno (Minneapolis: University of Minnesota Press, 2001), xvi.

48. David Silver, "Looking Backwards, Looking Forward: Cyberculture Studies 1990–2000," in *Web.Studies: Rewiring Media Studies for the Digital Age*, ed. David Gauntlett (London: Arnold, 2000), 24.

49. Silver, "Looking Backwards, Looking Forward," 24–25.

50. Raymond Williams, *The Country and the City* (New York: Oxford University Press, 1975), 296.

51. Richard Sennett, *The Hidden Injuries of Class* (Cambridge: Cambridge University Press, 1972), 176.

52. Alice Mary Hilton, "Cybernetics and Cybernation," *Science Teacher* 40, no. 2 (1973): 40.

53. Hilton, "Cybernetics and Cybernation," 40.

54. Hilton, "Cybernetics and Cybernation," 40.

4. Distortion

1. Samuel R. Delany, *Starboard Wine: More Notes on the Language of Science Fiction* (Middletown, CT: Wesleyan University Press, 2012), 165.

2. Ursula K. Le Guin, *Conversations with Ursula K. Le Guin*, ed. Carl Freedman (Jackson: University Press of Mississippi, 2008), 3.

3. Alexis Shotwell, *Against Purity: Living Ethically in Compromised Times* (Minneapolis: University of Minnesota Press, 2016), 186.

4. Martin Luther King, Jr., "Remaining Awake through a Great Revolution," in *A Knock at Midnight: Inspiration from the Great Sermons of Reverend Martin Luther King, Jr.*, ed. Clayborne Carson and Peter Holloran (New York: Warner Books, 2000), 206–207.

5. King, "Remaining Awake through a Great Revolution," 207–208.

6. Martin Luther King, Jr., *Where Do We Go from Here: Chaos or Community?* (Boston: Beacon Press, 1968), 179.

7. King, *Where Do We Go from Here?*, 201.

8. Samuel R. Delany, *The American Shore: Meditations on a Tale of Science Fiction by Thomas M. Disch* (Middletown, CT: Wesleyan University Press, 2014), 36.

9. N. Katherine Hayles, *How We Became Posthuman: Virtual Bodies in Cybernetics, Literature, and Informatics* (Chicago: University of Chicago Press, 2008), 84.

10. Madhu Dubey, *Signs and Cities: Black Literary Postmodernism* (Chicago: University of Chicago Press, 2003), 187.

11. Delany, *American Shore*, 1.

12. Melinda Gates, "Creating a Brotherhood" (commencement address, Duke University, Durham, NC, May 12, 2013), *Duke Today*, July 23, 2016, http://today.duke.edu/2013/05/gatestalk.

13. Philip José Farmer, "Riders of the Purple Wage," in *Dangerous Visions*, ed. Harlan Ellison (New York: Doubleday 1967), 132.

14. Philip José Farmer, "REAP: The Baycon Guest-of-Honor Speech," *Science Fiction Review* 28 (1968): 5.

15. Farmer, "REAP," 14.

16. Farmer, "REAP," 16. Notably, the "Fertile Void" is a formula devised by Paul Goodman, in fiction and in theory, before its extension into gestalt therapeutic

practice. It refers to a political and aesthetic space of pain and possibility, as described in Goodman's novel *The Empire City*: "The Fertile Void is yielding up its wonders one by one, mouthaches and growling rage, and a flood of boiling tears. Do you know?" On this and other aspects of Goodman as they inflect technological and political thinking, see my "Talking Politics in the Fertile Void," in *What Lies Between: Void Aesthetics and Postwar Post-Politics* (2015).

17. Leslie Fiedler, "Notes on Philip José Farmer," in *The Devil Gets His Due: The Uncollected Essays of Leslie Fiedler*, ed. Samuele F. S. Pardini (Berkeley: Counterpoint, 2008), 233–234.

18. Delany, *Starboard Wine*, 195.

19. Philip José Farmer, "Blueprint for Free Beer," *Knight*, July 1967, 93.

20. Shulamith Firestone, *The Dialectic of Sex: The Case for Feminist Revolution* (New York: Farrar, Straus, and Giroux, 1970), 184.

21. Susana Paasonen, "From Cybernation to Feminization: Firestone and Cyber-feminism," in *Further Adventures of the Dialectic of Sex: Critical Essays on Shulamith Firestone*, ed. Mandy Merck and Stella Sandford (London: Palgrave MacMillan, 2010), 74–75.

22. Paasonen, "From Cybernation to Feminization," 75.

23. Nina Power, "Toward a Cybernetic Communism: The Technology of the Anti-Family," in Merck and Sandford, *Further Adventures of the Dialectic of Sex,* 144.

24. Power, "Toward a Cybernetic Communism," 160.

25. Power, "Toward a Cybernetic Communism," 160.

26. Joanna Russ, Books, *Fantasy & Science Fiction*, November 1971, 18. *Skylark of Valeron* is a science-fiction novel by Edward E. Smith, first serialized in 1934.

27. Russ, Books (1971), 19.

28. Joanna Russ, "The Wearing Out of Genre Materials," *College English* 33, no. 1 (1971): 54.

29. Russ, "Wearing Out of Genre Materials," 54.

30. Israel Shenker, "Language Forum Hears Protests," *New York Times*, December 28, 1968, 18.

31. Shenker, "Language Forum Hears Protests," 18.

32. On the evolution of the Marxist Literary Group from the Radical Caucus at this event, for example, see Sean Homer, *Fredric Jameson: Marxism, Hermeneutics, Postmodernism* (28–29); on literary women's studies before and after, see chapter 9 of Florence Howe's autobiography, *A Life in Motion: A Memoir*. See also David Shumway, "The Sixties, the New Left, and the Emergence of Cultural Studies in the United States."

33. Delany, *American Shore*, 1.

34. Frederick Crews, "Do Literary Studies Have an Ideology?," *PMLA* 85, no. 3 (1970): 424.

35. H. Bruce Franklin, "The Teaching of Literature in the Highest Academies of the Empire," *College English* 31, no. 6 (1970): 549.

36. Alan C. Purves, "Life, Death, and the Humanities," *College English* 31, no. 6 (1970): 559.

37. Joanna Russ, "SF and Technology as Mystification," *Science Fiction Studies* 5, no. 3 (1978): 250. Russ's emphasis.

38. Russ, "SF and Technology as Mystification," 259.

39. H. Bruce Franklin, *Future Perfect: American Science Fiction of the Nineteenth Century* (Oxford: Oxford University Press, 1966), 143.

40. Franklin, *Future Perfect*, xii.

41. Teresa de Lauretis, introduction to *The Technological Imagination: Theories and Fictions*, ed. Teresa de Lauretis et al. (Madison, WI: Coda Press, 1980), 139.

42. Donna Haraway, *Simians, Cyborgs, and Women: The Reinvention of Women* (New York: Routledge, 1991), 149.

43. James Laughlin, introduction to *New Directions* 18 (1964): ix.

44. Laughlin, introduction, x.

45. Thomas Merton, *The Hidden Ground of Love: Letters*, ed. William H. Shannon (New York: Farrar, Straus, Giroux, 1985), 216.

46. Thomas Merton, *Turning toward the World: The Pivotal Years*, ed. Victor A. Kramer (San Francisco: Harper San Francisco, 1998), 207.

47. The phrase is Merton's, from the title of his 1968 essay "War and the Crisis of Language."

48. Thomas Merton, *Cables to the Ace: Familiar Liturgies of Misunderstanding* (New York: New Directions, 1968), 5–6.

49. Merton, *Cables to the Ace*, 2.

50. For Benjamin, in his discussion of a prior technological modernity, this parrying is a way by which a potentially traumatic shock may be delimited and assimilated into isolated experience (*Erlebnis*) rather than being left to fester and repeat its harm to experience over a long term (*Erfahrung*). (Walter Benjamin, "On Some Motifs in Baudelaire," in *Selected Writings*, vol. 4, *1938–1940*, ed. Howard Eiland and Michael W. Jennings, trans. Edmund Jephcott [Cambridge, MA: Harvard Belknap, 2003], 318–319.)

5. Revolutionary Suicide

1. Huey P. Newton, *Revolutionary Suicide* (New York: Penguin, 2009), 3.

2. Joanna Russ, *We Who Are About To . . .* (Middletown, CT: Wesleyan University Press, 2005), 34.

3. In Russ, *We Who Are About To . . .*, x.

4. Russ, *We Who Are About To . . .*, 37.

5. Rebekah Sheldon, *The Child to Come: Life after the Human Catastrophe* (Minneapolis: University of Minnesota Press, 2016), 83.

6. Russ, *We Who Are About To . . .*, 2.

7. Russ, *We Who Are About To . . .*, 104.

8. Russ, *We Who Are About To . . .*, 95.

9. Russ, *We Who Are About To . . .*, 117.

10. Paul Metcalf, *Collected Works*, vol. 1, *1956–1976* (Minneapolis: Coffee House Press, 1996), 413.

11. Metcalf, *Collected Works*, 423–424.

12. Metcalf, *Collected Works*, 422.

13. Frederick Douglass, *Autobiographies* (New York: Library of America, 1994), 404.

14. William Wells Brown, *Clotel; or, The President's Daughter* (New York: Penguin, 2004), 183.

15. C.L.R. James, *Black Jacobins* (New York: Vintage, 1989), 15.

16. Orlando Patterson, *The Sociology of Slavery: An Analysis of the Origins, Development, and Structure of Negro Slave Society in Jamaica* (Teaneck, NJ: Farleigh Dickinson University Press, 1967), 264.

17. Hortense J. Spillers, *Black, White, and in Color: Essays on American Literature and Culture* (Chicago: University of Chicago Press, 2003), 183.

18. Exemplary in this regard is Terri L. Snyder's *The Power to Die: Slavery and Suicide in British North America* (Chicago: University of Chicago Press, 2015), which relies on many of the same archival sources that occupy Metcalf. Snyder notes that many narrations of slave suicide were motivated by the material interests of their narrators, yet also that "in radically reframing the political implications and cultural meanings of suicide and slavery for early Americans, the power to die was nothing short of revolutionary" (156).

19. Metcalf, *Collected Works*, 440.

20. Hayden Carruth, "Ludd, Efik, and Orca," *Harper's*, January 1977, 12.

21. Paul Metcalf, *Where Do You Put the Horse?* (New York: W.W. Norton 1986), 16.

22. Thomas Pynchon, "Is It O.K. to Be a Luddite?," *New York Times Book Review*, October 28, 1984, 40–41.

23. Friedrich Kittler, "Media and Drugs in Pynchon's Second World War," in *Reading Matters: Narrative in the New Media Ecology*, ed. Joseph Tabbi and Michael Wutz, trans. Michael Wutz and Geoffrey Winthrop-Young (Ithaca, NY: Cornell University Press, 1997), 163–164.

24. Kathleen Fitzpatrick, *The Anxiety of Obsolescence: The American Novel in the Age of Television* (Nashville: Vanderbilt University Press, 2006), 77–78.

25. Toni Morrison, *Sula* (New York: Alfred A. Knopf, 1973), 14.

26. Audre Lorde, "*Sula* / A Review." *Amazon Quarterly* 2, no. 3 (1974): 28.

27. Morrison, *Sula*, 7.

28. Houston A. Baker, Jr. *Workings of the Spirit: The Poetics of Afro-American Women's Writing* (Chicago: University of Chicago Press, 1991), 138.

29. Baker, *Workings of the Spirit*, 156.

30. In Shiva Naipaul, *Journey to Nowhere: A New World Tragedy* (New York: Penguin, 1982), 158.

31. In Naipaul, *Journey to Nowhere*, 287–288.

32. In Naipaul, *Journey to Nowhere*, 289.

6. Liberation Technology

1. Richard Barbrook and Andy Cameron, "The California Ideology," *Science as Culture* 6, no. 1 (1996): 44–72.

2. In Kenneth Rosen, ed., *Voices of the Rainbow: Contemporary Poetry by Native Americans* (New York: Viking Press, 1975), 80.

3. In Rosen, *Voices of the Rainbow*, 80.

4. In Rosen, *Voices of the Rainbow*, 80.

5. John Mohawk, "Our Strategy for Survival," in *Basic Call to Consciousness* (Summertown, TN: Native Voices, 1978), 122.

6. Mohawk, "Our Strategy for Survival," 121.

7. Mohawk, "Our Strategy for Survival," 122.

8. John Mohawk, "Technology Is the Enemy," *Akwesasne Notes* 1979, 21.

9. Mohawk, "Technology Is the Enemy," 21.

10. Richard A. Falk, "Anarchism and World Order," *Nomos* 19 (1978): 65–66.

11. Richard A. Falk, "Anarchism and World Order," 83.

12. E.F. Schumacher, *Small Is Beautiful: Economics as If People Mattered* (New York: Harper & Row, 1973), 167.

13. Schumacher, *Small Is Beautiful*, 170.

14. Gustavo Gutiérrez, *A Theology of Liberation* (Maryknoll, NY: Orbis, 1973), 75–76.

15. Mohawk, "Our Strategy for Survival," 117.

16. Langdon Winner, "Peter Pan in Cyberspace: *Wired* Magazine's Political Vision," *Educom Review,* May/June 1995, 19.

17. Ayn Rand, "The Anti-Industrial Revolution," *Objectivist* 10 (1971): 7.

18. Rand, "Anti-Industrial Revolution," 3–4.

19. In Veronika Bennholt-Thomsen and Maria Mies, *The Subsistence Perspective: Beyond the Globalised Economy* (London: Zed Books, 2000), 25.

20. Carol Hill, *Subsistence, U.S.A.* (New York: Holt, Rinehart & Winston, 1973), 7.

21. Yorick A. Wilks, Brian M. Slator, and Louise M. Guthrie, *Electric Words: Dictionaries, Computers, and Meanings* (Cambridge, MA: MIT Press, 1996), 81.

22. Carter Revard, "On the Computability of Certain Monsters in Noah's Ark: Using Computers to Study *Webster's Seventh New Collegiate Dictionary* and *The New Merriam-Webster Pocket Dictionary*," in *Proceedings of the 1968 ACM [Association for Computing Machinery] National Conference* (New York: ACM, 1968): 812.

23. Carter Revard, "How to Make a N.U.D.E. (New Utopian Dictionary of English)," *Annals of the New York Academy of Sciences* 211, no. 1 (1973): 95.

24. Larry Diamond, "Liberation Technology," *Journal of Democracy* 21, no. 3 (2010): 70. It is worth adding that Diamond, in his youth, endeavored to be a moderating force in the student-led political actions of the late 1960s. Indeed, during the antiwar occupation of the Stanford Computer Center, discussed in chapter 1, Diamond was seen as a key obstruction. Along with Bob Grant, another member of the Stanford community, Diamond was perceived to have interfered with the activities of the campus Left. Many years passed between these events and Diamond's later expressions of enthusiasm for the ostensibly liberatory value of corporate-owned social media, and surely one may have nothing to do with the other. Yet it bears reporting that on the day of the occupation, Grant and Diamond discouraged the action. Franklin responded in his noonday speech: "There were some hot emotions at the beginning when Bob Grant and Larry Diamond tried to subvert what we were doing. And I think a lot of people misunderstood where things were and what was coming down. Because they believed that they were very sincere people and so forth . . . [but] the fact of the matter is that a lot of us were doing precinct work out in the community in 1964, and at that time we were opposed by the Bob Grants and Larry Diamonds of the world." 118 Cong. Rec., no. 4 (1972), 4805.

25. Jarrett Martineau, "Rhythms of Change: Mobilizing Decolonial Consciousness, Indigenous Resurgence and the Idle No More Movement," in *More Will Sing Their Way to Freedom: Indigenous Resistance and Resurgence*, ed. Elaine Coburn (Halifax, NS: Fernwood, 2015), 245.

7. Thanatopography

1. Alexander R. Galloway, with David M. Berry, "A Network Is a Network Is a Network: Reflections on the Computational and the Societies of Control," *Theory Culture & Society* 33, no. 4 (2016): 160.

2. Galloway, with Berry, "Network Is a Network," 160.

3. Norbert Wiener, *Cybernetics, or Control and Communication in the Animal and the Machine* (Cambridge, MA: MIT Press, 1948), 28–29.

4. Nellie Wong, "On the Crevices of Anger," *Conditions* 1978, 52–53.

5. Wong, "On the Crevices of Anger," 53, 55.

6. Wong, "On the Crevices of Anger," 55.

7. Dominick LaCapra, *History and Memory after Auschwitz* (Ithaca, NY: Cornell University Press, 1998), 47.

8. Kurihara Sadako, *When We Say "Hiroshima": Selected Poems* (Ann Arbor: University of Michigan Press, 1999), 54.

9. Ran Zwigenberg, "Never Again: Hiroshima, Auschwitz and the Politics of Commemoration," *Asia-Pacific Journal* 13, no. 3 (2015): 16.

10. Raphael Sassower, *Technoscientific Angst: Ethics and Responsibility* (Minneapolis: University of Minnesota Press, 1997), 64.

11. In Cordelia Candelaria, ed., *Multiethnic Literature of the United States: Critical Introductions and Classroom Resources*. Boulder, CO: Multiethnic Literature Project, 1989, 67–68.

12. In Candelaria, *Multiethnic Literature of the United States*, 67.

13. Tung-Hui Hu, *A Prehistory of the Cloud* (Cambridge, MA: MIT Press, 2015), 18.

14. Robert Jay Lifton, "Absurd Technological Death," in *Crimes of War: A Legal, Political-Documentary, and Psychological Inquiry into the Responsibility of Leaders, Citizens, and Soldiers for Criminal Acts in Wars*, ed. Richard A. Falk, Gabriel Kolko, and Robert Jay Lifton (New York: Random House, 1971), 26.

15. Derek Walcott, "The Muse of History," in *Is Massa Day Dead?*, ed. Orde Coombs (New York: Doubleday Anchor, 1974), 9.

16. Steve Lichtgarden, "Con Ed Crematorium," *East Village Other* 1 (1967): 5.

17. "Not Good Germans," *Fifth Estate*, April 22–28, 1971, 3.

18. Theodor W. Adorno, "Cultural Criticism and Society," in *Prisms* (Cambridge, MA: MIT Press, 1967), 34.

19. In Joseph Bruchac, ed., *The Next World: Poems by 32 Third World Americans* (Trumansburg, NY: Crossing Press, 1978), 202.

20. Henry Dumas, *Knees of a Natural Man: The Selected Poems*, ed. Eugene B. Redmond (New York: Thunder's Mouth Press, 1989), 70.

21. Ricardo Sanchez, "calles y callejones & memories," *A: A Journal of Contemporary Literature* 4, no. 1 (1979): 12–13.

22. Giorgio Agamben, *Remnants of Auschwitz: The Witness and the Archive*, trans. Daniel Heller-Roazen (New York: Zone Books, 2000), 101. Agamben's emphasis.

23. Agamben, *Remnants of Auschwitz*, 77.

24. Armand Mattelart, *Networking the World, 1794–2000*, trans. Liz Carey-Libbrecht and James A. Cohen (Minneapolis: University of Minnesota Press, 2000), 120.

25. Krzysztof Ziarek, "The Way of the World, or the Critique of Cosmo-Technics," *Parallax* 16, no. 4 (2010): 3.

26. Jean-Luc Nancy, *After Fukushima: The Equivalence of Catastrophes*, trans. Charlotte Mandell (New York: Fordham University Press, 2014), 12.

27. Nancy, *After Fukushima*, 12–13.

28. Galloway and Berry, "Network Is a Network," 160.

29. Stanley Elkin, "The Conventional Wisdom," *American Review* 26 (1977): 94.

30. Elkin, "Conventional Wisdom," 108.

31. Elkin, "Conventional Wisdom," 108–109.

32. In Walter Abish, *In the Future Perfect* (New York: New Directions, 1977), 1.

33. Abish, *In the Future Perfect*, 4.

34. Abish, *In the Future Perfect*, 9.

35. Abish, *In the Future Perfect*, 4.

36. Harry Mathews and Georges Perec, "Roussel and Venice: Outline of a Melancholic Geography," in *Immeasurable Distances: The Collected Essays*, by Harry Mathews (Venice, CA: Lapis Press, 1991), 90.

37. Mathews and Perec, "Roussel and Venice," 91.

38. Warren F. Motte, "Eradications," *Romance Notes* 29, no. 1 (1988): 36.

39. J.-B. Pontalis, *Love of Beginnings*, trans. James Greene with Marie-Christine Réguis (London: Free Association Books, 1993), 145.

40. Mathews and Perec, "Roussel and Venice," 106.

41. Mathews and Perec, "Roussel and Venice," 88.

42. Shoshana Felman and Dori Laub, *Testimony: Crises of Witnessing in Literature, Psychoanalysis, and History* (New York: Routledge, 1992), 53.

43. Daniel Tiffany, *Toy Medium: Materialism and the Modern Lyric* (Berkeley: University of California Press, 2000), 226–227.

44. Alexander Kuo, "New Letters from Hiroshima," *Yardbird Reader* 3 (1974): 43.

45. Akira Mizuta Lippitt, *Atomic Light (Shadow Optics)* (Minneapolis: University of Minnesota Press, 2005), 86.

46. Kuo, "New Letters from Hiroshima," 43–44.

47. In Fay Chiang et al., eds. *American Born and Foreign: An Anthology of Asian American Poetry* (Bronx: Sunbury, 1979), 82.

48. In Chiang, *American Born and Foreign*, 83.

49. Nanao Sakaki, "Memorandum," *Bombay Gin* (1979): 116.

50. Sakaki, "Memorandum," 117.

51. Paula Gunn Allen, "In a Tavern at White Sands," *A: A Journal of Contemporary Literature* (1978): 30.

52. In Candelaria, *Multiethnic Literature of the United States*, 68.

Conclusion: American Carnage and Technologies of Tomorrow

1. Donald J. Trump (with Steve Bannon and Steven Miller), "The Inaugural Address" (speech, Washington, DC, January 20, 2017), https://www.whitehouse.gov/briefings-statements/the-inaugural-address/.

2. Trump, "Inaugural Address."

3. Antonio Gramsci, *Selections from the Prison Notebooks*, ed. and trans. Quintin Hoare and Geoffrey Nowell-Smith (New York: International Publishers, 1978), 478.

4. Gramsci, *Selections from the Prison Notebooks*, 328.

5. Houston A. Baker, Jr., *Workings of the Spirit: The Poetics of Afro-American Women's Writing* (Chicago: University of Chicago Press, 1991), 156.

6. Baker, *Workings of the Spirit*, 138.

7. Teju Cole, "The Superhero Photographs of the Black Lives Matter Movement," *New York Times Magazine*, July 26, 2016, 16.

8. R.H. Lossin, "On Sabotage," *Politics/Letters*, June 27, 2016, http://politicss-lashletters.org/on-sabotage/.

9. Lossin, "On Sabotage."

10. Richard Kauzlarich, "House Energy and Commerce Hearing Highlights Critical Need to Invest in America's Infrastructure," *The Hill*, February 22, 2017, https://thehill.com/blogs/congress-blog/energy-environment/320547-house-energy-and-commerce-hearing-highlights-critical.

11. Ayn Rand, "The Anti-Industrial Revolution," *Objectivist* 10 (1971): 7.

12. Lorenna Bravebull Allard, "Why Do We Punish Dakota Pipeline Protesters but Exonerate the Bundys?" *Guardian*, November 2, 2016, https://www.theguardian.com/commentisfree/2016/nov/02/dakota-pipeline-protest-bundy-militia.

13. Jonathan Beller, "Prosthetics of Whiteness: Drone Psychosis," *The Message Is Murder: Substrates of Computational Capital* (London: Pluto Press, 2018), 144.

14. In Carl Mitcham and Robert Mackey, eds., *Philosophy and Technology: Readings in the Philosophical Problems of Technology* (New York: Free Press, 1972), 130–134.

15. Babette Babich, "Angels, the Space of Time, and Apocalyptic Blindness: On Günther Anders' Endzeit-Endtime," *Etica & Politica / Ethics & Politics* 15, no. 2 (2013): 155.

16. In Mitcham and Mackey, *Philosophy and Technology*, 134

17. Charlotte Bunch, "The Reform Tool Kit," *Quest: A Feminist Quarterly* 1, no. 1 (1974): 45.

18. Bunch, "Reform Tool Kit," 45–46.

19. Raphael Sassower, *Technoscientific Angst: Ethics and Responsibility* (Minneapolis: University of Minnesota Press, 1997), 97–99.

20. Sarah Kember and Joanna Zylinska, *Life after New Media: Mediation as a Vital Process* (Cambridge, MA: MIT Press, 2012), xi.

21. Andrew McAfee and Erik Brynjolfsson, *Machine, Platform, Crowd: Harnessing Our Digital Future* (New York: W.W. Norton, 2017), 24.

22. McAfee and Brynjolfsson, *Machine, Platform, Crowd*, 334.

23. McAfee and Brynjolfsson, *Machine, Platform, Crowd*, 334.

24. James Boggs, Grace Lee Boggs, Freddy Paine, and Lyman Paine, *Conversations in Maine: Exploring Our Nation's Future* (Boston: South End Press, 1978), 100.

25. Boggs et al., *Conversations in Maine*, 100.

26. Boggs et al., *Conversations in Maine*, 48.

WORKS CITED

118 Congressional Record, no. 4. 1972.

Abish, Walter. *In the Future Perfect*. New York: New Directions, 1977.

Abraham, Nicolas, and Mária Török. *The Shell and the Kernel*. Volume 1. Edited by Nicholas T. Rand. Chicago: University of Chicago Press, 1994.

Ad Hoc Committee. "The Triple Revolution: Cybernation—Weaponry—Human Rights." In *Seeds of Liberation*. Edited by Paul Goodman. New York: George Braziller, 1964.

Adorno, Theodor W. "Cultural Criticism and Society." In *Prisms*, 17–34. Cambridge, MA: MIT Press, 1967.

Adorno, Theodor W. "Education after Auschwitz." In *Critical Models: Interventions and Catchwords*, translated by Henry W. Pickford, 191–204. New York: Columbia University Press, 1998.

Adorno, Theodor W. "Late Capitalism or Industrial Society? The Fundamental Question of the Present Structure of Society." In *Can One Live after Auschwitz? A Philosophical Reader*, 111–125. Palo Alto: Stanford University Press, 2003.

Agamben, Giorgio. *Remnants of Auschwitz: The Witness and the Archive*. Translated by Daniel Heller-Roazen. New York: Zone Books, 2000.

Ahmed, Sara. *Living a Feminist Life*. Durham, NC: Duke University Press, 2017.

Allen, Paula Gunn. "In a Tavern at White Sands." *A: A Journal of Contemporary Literature* (1978): 30.

Arendt, Hannah. *The Human Condition*. 2nd ed. Chicago: University of Chicago Press, 1998.

Arendt, Hannah. *The Portable Hannah Arendt*. Edited by Peter Baehr. New York: Penguin, 2000.

Ashbery, John. *Selected Poems*. New York: Viking Penguin, 1985.

Babich, Babette. "Angels, the Space of Time, and Apocalyptic Blindness: On Günther Anders' Endzeit-Endtime." *Etica & Politica / Ethics & Politics* 15, no. 2 (2013): 144–174.

Baker, Houston A., Jr. *Workings of the Spirit: The Poetics of Afro-American Women's Writing*. Chicago: University of Chicago Press, 1991.

Bakhtin, M.M. *The Dialogic Imagination: Four Essays*. Edited by Michael Holquist. Translated by Caryl Emerson and Michael Holquist. Austin: University of Texas Press, 1981.

Barbrook, Richard, and Andy Cameron. "The California Ideology." *Science as Culture* 6, no. 1 (1996): 44–72.

Bates, Dorothy. "The Machine I Hate the Most." *Avant Garde*, 1971, 22–23.

Beller, Jonathan. "Prosthetics of Whiteness: Drone Psychosis." In *The Message Is Murder: Substrates of Computational Capital*, 137–157. London: Pluto Press, 2018.

Benjamin, Walter. "On Some Motifs in Baudelaire." In *Selected Writings,* vol. 4, *1938–1940,* edited by Howard Eiland and Michael W. Jennings, 313–355. Translated by Edmund Jephcott. Cambridge, MA: Harvard Belknap, 2003.

Benjamin, Walter. *Origin of the German Trauerspiel.* Translated by Howard Eiland. Cambridge, MA: Harvard University Press, 2019.

Bennholt-Thomsen, Veronika, and Maria Mies. *The Subsistence Perspective: Beyond the Globalised Economy.* London: Zed Books, 2000.

Berlant, Lauren. *The Female Complaint: The Unfinished Business of Sentimentality in American Culture.* Durham, NC: Duke University Press, 2008.

Binfield, Kevin, ed. *Writings of the Luddites.* Baltimore: Johns Hopkins University Press, 2004.

Boggs, James, Grace Lee Boggs, Freddy Paine, and Lyman Paine. *Conversations in Maine: Exploring Our Nation's Future.* Boston: South End Press, 1978.

Bookchin, Murray. *Post-Scarcity Anarchism.* 2nd ed. Montreal: Black Rose Books, 1986.

Borgmann, Albert. "The Explanation of Technology." *Research in Philosophy and Technology* 1 (1978): 99–118.

Bové, Paul A. *Destructive Poetics: Heidegger and Modern American Poetry.* New York: Columbia University Press, 1980.

Brand, Stewart. *Whole Earth Catalog.* Menlo Park: Portola Institute, 1968.

Bravebull Allard, Lorenna. "Why Do We Punish Dakota Pipeline Protesters but Exonerate the Bundys?" *Guardian,* November 2, 2016. https://www.theguardian.com/commentisfree/2016/nov/02/dakota-pipeline-protest-bundy-militia.

Brick, Howard, and Christopher Phelps. *Radicals in America: The U.S. Left since the Second World War.* Vol. 1. Cambridge: Cambridge University Press, 2015.

Brown, William Wells. *Clotel; or, The President's Daughter.* New York: Penguin, 2004.

Bruchac, Joseph, ed. *The Next World: Poems by 32 Third World Americans.* Trumansburg, NY: Crossing Press, 1978.

Bunch, Charlotte. "The Reform Tool Kit." *Quest: A Feminist Quarterly* 1, no. 1 (1974): 37–51.

Bunge, Mario. "The Five Buds of Technophilosophy." *Technology in Society* 1 (1979): 67–74.

Burke, Kenneth. *Language as Symbolic Action.* Berkeley: University of California Press, 1966.

Burke, Kenneth. "Routine for a Stand-Up Comedian." In *Late Poems 1968–1993: Attitudinizings Verse-Wise, while Fending for One's Selph, and in a Style Somewhat Artificially Colloquial,* edited by Julie Whitaker and David Blakesley. Columbia: University of South Carolina Press, 2005.

Burke, Kenneth. "Why Satire, with a Plan for Writing One." *Michigan Quarterly Review* 13, no. 4 (1974): 303–337.

Candelaria, Cordelia, ed. *Multiethnic Literature of the United States: Critical Introductions and Classroom Resources.* Boulder: Multiethnic Literature Project, 1989.

Carruth, Hayden. "Ludd, Efik, and Orca." *Harper's,* January 1977, 12.

Chiang, Fay, et al., eds. *American Born and Foreign: An Anthology of Asian American Poetry.* Bronx: Sunbury, 1979.

Cole, Teju. "The Superhero Photographs of the Black Lives Matter Movement." *New York Times Magazine,* July 26, 2016, 16, 18–19.

"Constitution of UNESCO." November 16, 1945. UNESCO. http://portal.unesco. org/en/ev.php-URL_ID=15244&URL_DO=DO_TOPIC&URL_SECTION =201.html.

Crews, Frederick. "Do Literary Studies Have an Ideology?" *PMLA* 85, no. 3 (1970): 423–428.

Davis, Mike, et al. Quoted in Louis Proyect, "Robert Brenner, Vivek Chibber, and the 'Organization Question.'" *Louis Proyect: The Unrepentant Marxist* (blog). June 25, 2018. https://louisproyect.org/2018/06/25/robert-brenner-vivek-chibber-and-the-organization-question/.

de Kerckhove, Derrick. *The Skin of Culture: Investigating the New Electronic Reality.* Toronto: Somerville House, 1995.

Delany, Samuel R. "The *Algol* Interview: Samuel R. Delany." *Algol* 13 (1976): 16–20.

Delany, Samuel R. *The American Shore: Meditations on a Tale of Science Fiction by Thomas M. Disch.* Middletown, CT: Wesleyan University Press, 2014.

Delany, Samuel R. Introduction to *The Cosmic Rape and "To Marry Medusa,"* by Theodore Sturgeon, i–xxxiv. Boston: Gregg Press, 1977.

Delany, Samuel R. *Starboard Wine: More Notes on the Language of Science Fiction.* Middletown, CT: Wesleyan University Press, 2012.

Delany, Samuel R. *Triton: An Ambiguous Heterotopia.* New York: Bantam Books, 1976.

de Lauretis, Teresa. Introduction to *The Technological Imagination: Theories and Fictions,* edited by Teresa de Lauretis et al., 135–39. Madison, WI: Coda Press, 1980.

Diamond, Larry. "Liberation Technology." *Journal of Democracy* 21, no. 3 (2010): 69–83.

Dotson, Kristie. "How Is This Paper Philosophy?" *Comparative Philosophy* 3, no. 1 (2012): 3–29.

Douglass, Frederick. *Autobiographies.* New York: Library of America, 1994.

Dubey, Madhu. *Signs and Cities: Black Literary Postmodernism.* Chicago: University of Chicago Press, 2003.

Dumas, Henry. *Knees of a Natural Man: The Selected Poems.* Edited by Eugene B. Redmond. New York: Thunder's Mouth Press, 1989.

Dupuy, Jean-Pierre. "Myths of the Informational Society." In *The Myths of Information: Technology and Postindustrial Culture,* edited by Kathleen Woodward, 3–17. Madison, WI: Coda Press, 1980.

Elkin, Stanley. "The Conventional Wisdom." *American Review* 26 (1977): 79–109.

Falk, Richard A. "Anarchism and World Order." *Nomos* 19 (1978): 63–87.

Farmer, Philip José. "Blueprint for Free Beer." *Knight,* July 1967, 8–9, 84–85, 90–93.

Farmer, Philip José. "REAP: The Baycon Guest-of-Honor Speech." *Science Fiction Review* 28 (1968): 4–16.

Farmer, Philip José. "Riders of the Purple Wage." In *Dangerous Visions,* edited by Harlan Ellison, 62–134. New York: Doubleday, 1967.

Feenberg, Andrew. "The Political Economy of Social Space." In *The Myths of Information: Technology and Postindustrial Culture,* edited by Kathleen Woodward, 111–124. Madison, WI: Coda Press, 1980.

Felman, Shoshana, and Dori Laub. *Testimony: Crises of Witnessing in Literature, Psychoanalysis, and History.* New York: Routledge, 1992.

Fenton, David, and Gil Scott-Heron. "'The First Minute of a New Day': Music and Politics with Gil Scott-Heron." *Ann Arbor Sun*, March 14, 1975, 19, 21.

Ferguson, Roderick A. "Of Sensual Matters: On Audre Lorde's 'Poetry Is Not a Luxury' and 'Uses of the Erotic.'" *Women's Studies Quarterly* 40, no. 3/4 (2012): 295–300.

Fiedler, Leslie. "Notes on Philip José Farmer." In *The Devil Gets His Due: The Uncollected Essays of Leslie Fiedler*, edited by Samuele F. S. Pardini, 230–235. Berkeley: Counterpoint, 2008.

Firestone, Shulamith. *The Dialectic of Sex: The Case for Feminist Revolution*. New York: Farrar, Straus, and Giroux, 1970.

Fitzpatrick, Kathleen. *The Anxiety of Obsolescence: The American Novel in the Age of Television*. Nashville: Vanderbilt University Press, 2006.

Foucault, Michel. *The Archaeology of Knowledge and the Discourse on Language*. Translated by A.M. Sheridan Smith. New York: Pantheon Books, 1972.

Franklin, H. Bruce. *Future Perfect: American Science Fiction of the Nineteenth Century*. Oxford: Oxford University Press, 1966.

Franklin, H. Bruce. *Robert A. Heinlein: America as Science Fiction*. Oxford: Oxford University Press, 1980.

Franklin, H. Bruce. "The Teaching of Literature in the Highest Academies of the Empire." *College English* 31, no. 6 (1970): 548–557.

Freeman, Elizabeth A. *Time Binds: Queer Temporalities, Queer Histories*. Durham, NC: Duke University Press, 2010.

Fuller, R. Buckminster. "The Comprehensive Man." *Northwest Review* 2, no. 2 (1959): 23–33.

Fuller, R. Buckminster. "Man's Total Communication System." *Evergreen Review* 14, no. 83 (1970): 39–41, 59–65.

Fuller, R. Buckminster. *Operating Manual for Spaceship Earth*. Carbondale: Southern Illinois University Press, 1968.

Galloway, Alexander R. With David M. Berry. "A Network Is a Network Is a Network: Reflections on the Computational and the Societies of Control." *Theory Culture & Society* 33, no. 4 (2016): 151–172.

Gates, Melinda. "Creating a Brotherhood." Commencement address, Duke University, Durham, NC, May 12, 2013. *Duke Today*, July 23, 2016. http://today.duke.edu/2013/05/gatestalk.

Glissant, Édouard. *Caribbean Discourse: Selected Essays*. Translated by Michael Dash. Charlottesville: University of Virginia Press, 1989.

Glissant, Édouard. *Poetic Intention*. Translated by Nathalie Stephens. Callicoon, NY: Nightboat Books, 2010.

Goodman, Paul. "The Trouble with Today's Technology: A Social Critic's View." *Innovation*, June 1969, 38–47.

Gramsci, Antonio. *Selections from the Prison Notebooks*. Edited and translated by Quintin Hoare and Geoffrey Nowell-Smith. New York: International Publishers, 1978.

Gutiérrez, Gustavo. *A Theology of Liberation*. Maryknoll, NY: Orbis, 1973.

Hannay, N. Bruce, and Robert E. McGinn. "The Anatomy of Modern Technology: Prolegomenon to an Improved Public Policy for the Social Management of Technology." *Daedalus* 109, no. 1 (1980): 25–53.

Haraway, Donna. *Simians, Cyborgs, and Women: The Reinvention of Women.* New York: Routledge, 1991.

Hardin, Garrett. "Living on a Lifeboat." *BioScience* 24, no. 10 (1974): 561–568.

Hayles, N. Katherine. *How We Became Posthuman: Virtual Bodies in Cybernetics, Literature, and Informatics.* Chicago: University of Chicago Press, 2008.

Heidegger, Martin. *The Question concerning Technology and Other Essays.* Translated by William Lovitt. New York: Harper and Row, 1977.

Hicks, Heather. *The Culture of Soft Work: Labor, Gender, and Race in Postmodern American Narrative.* New York: Palgrave MacMillan, 2009.

Hill, Carol. *Subsistence, U.S.A.* New York: Holt, Rinehart & Winston, 1973.

Hilton, Alice Mary. "Cybernetics and Cybernation." *Science Teacher* 40, no. 2 (1973): 34–40.

Hilton, Alice Mary. "Cybernetics and the Future." Interview with Studs Terkel. Chicago, WFMT Radio Network, July 12, 1965.

Hilton, Alice Mary, ed. *The Evolving Society: The Proceedings of the First Annual Conference on the Cybercultural Revolution.* New York: Institute for Cybercultural Research, 1966.

Hilton, Alice Mary. "Hilton's Law." *Improving College and University Teaching* 16, no.2 (1968): 149.

Hilton, Alice Mary. "Individual Responsibility in a Cybercultural Society." *Quest* 5 (1965): 37–47.

Hilton, Alice Mary. *Logic, Computing Machines, and Automation.* Washington, DC: Spartan Books, 1963.

Hobsbawm, Eric J. "The Machine Breakers." *Past & Present* 1 (1952): 57–70.

Hodel, Mike. "A Talk with Philip K. Dick." Interview with Philip K. Dick. North Hollywood, KPFK-FM Radio Network, June 26, 1976.

Hu, Tung-Hui. *A Prehistory of the Cloud.* Cambridge, MA: MIT Press, 2015.

Hui, Yuk. "On Cosmotechnics: For a Renewed Relation between Technology and Nature in the Anthropocene." *Techné: Research in Philosophy and Technology* 21, no. 2/3 (2017): 319–341.

Ihde, Don. *Technics and Praxis.* Dordrecht: D. Reidel, 1979.

Jackson, George. *Soledad Brother: The Prison Letters of George Jackson.* Edited by Jonathan Jackson, Jr. Chicago: Lawrence Hill Books, 1994.

James, C.L.R. *The Black Jacobins.* New York: Vintage, 1989.

Jameson, Fredric. "World-Reduction in Le Guin: The Emergence of Utopian Narrative." *Science Fiction Studies* 2, no. 3 (1975): 221–230.

John, Jipson, Jitheesh P.M., and David Harvey. "'The Neoliberal Project Is Alive but Has Lost Its Legitimacy': David Harvey." *The Wire*, February 9, 2019. https://thewire.in/economy/david-harvey-marxist-scholar-neo-liberalism.

Kauzlarich, Richard. "House Energy and Commerce Hearing Highlights Critical Need to Invest in America's Infrastructure." *The Hill*, February 22, 2017. https://thehill.com/blogs/congress-blog/energy-environment/320547-house-energy-and-commerce-hearing-highlights-critical.

Kember, Sarah, and Joanna Zylinska. *Life after New Media: Mediation as a Vital Process.* Cambridge, MA: MIT Press, 2012.

King, Martin Luther, Jr. "Remaining Awake through a Great Revolution." In *A Knock at Midnight: Inspiration from the Great Sermons of Reverend Martin Luther King, Jr.*, edited by Clayborne Carson and Peter Holloran, 201–224. New York: Warner Books, 2000.

King, Martin Luther, Jr. *Where Do We Go from Here: Chaos or Community?* Boston: Beacon Press, 1968.

Kittler, Friedrich. "Media and Drugs in Pynchon's Second World War." In *Reading Matters: Narrative in the New Media Ecology*, edited by Joseph Tabbi and Michael Wutz, 157–172. Translated by Michael Wutz and Geoffrey Winthrop-Young. Ithaca, NY: Cornell University Press, 1997.

Kuo, Alexander. "New Letters from Hiroshima." *Yardbird Reader* 3 (1974): 40–44.

LaCapra, Dominick. *History and Memory after Auschwitz*. Ithaca, NY: Cornell University Press, 1998.

Laughlin, James. Introduction. *New Directions* 18 (1964): ix–xii.

Le Guin, Ursula K. "The Algol Interview: Ursula K. Le Guin." *Algol* 12, no. 2 (1975): 7–10.

Le Guin, Ursula K. "The Child and the Shadow." *Quarterly Journal of the Library of Congress* 32, no. 2 (1975): 139–148.

Le Guin, Ursula K. *Conversations with Ursula K. Le Guin*. Edited by Carl Freedman. Jackson: University Press of Mississippi, 2008.

Le Guin, Ursula K. *The Dispossessed: An Ambiguous Utopia*. New York: Perennial Classics, 1974.

Lévy, Pierre. *Cyberculture*. Translated by Robert Bononno. Minneapolis: University of Minnesota Press, 2001.

Lichtgarden, Steve. "Con Ed Crematorium." *East Village Other* 1 (1967): 5.

Lifton, Robert Jay. "Absurd Technological Death." In *Crimes of War: A Legal, Political-Documentary, and Psychological Inquiry into the Responsibility of Leaders, Citizens, and Soldiers for Criminal Acts in Wars*, edited by Richard A. Falk, Gabriel Kolko, and Robert Jay Lifton. New York: Random House, 1971.

Lippitt, Akira Mizuta. *Atomic Light (Shadow Optics)*. Minneapolis: University of Minnesota Press, 2005.

Litwack, Michael. "Racial Technics: Media and Machines in the Long Civil Rights Era." PhD diss., Brown University, 2016.

Lorde, Audre. *From a Land Where Other People Live*. Detroit: Broadside Press, 1973.

Lorde, Audre. *Sister Outsider: Essays and Speeches*. Trumansburg, NY: Crossing Press, 1984.

Lorde, Audre. "*Sula*/A Review." *Amazon Quarterly* 2, no. 3 (1974): 28–30.

Lossin, R.H. "On Sabotage." *Politics/Letters*, June 27, 2016. http://politicsslashletters.org/on-sabotage/.

Lyotard, Jean-François. *The Postmodern Condition: A Report on Knowledge*. Translated by Geoff Bennington and Brian Massumi. Minneapolis: University of Minnesota Press, 1984.

MacLeish, Archibald. "A Reflection: Riders on the Earth Together, Brothers in Eternal Cold." *New York Times*, December 25, 1968, A1.

Marcuse, Herbert. "The End of Utopia." *Ramparts*, April 1970, 28–35.

Martineau, Jarrett. "Rhythms of Change: Mobilizing Decolonial Consciousness, Indigenous Resurgence and the Idle No More Movement." In *More Will Sing Their Way to Freedom: Indigenous Resistance and Resurgence*, edited by Elaine Coburn, 229–254. Halifax, NS: Fernwood, 2015.

Marx, Karl. *Capital: A Critique of Political Economy*. Vol. 1. Translated by Ben Fowkes. New York: Penguin Books, 1990.

Marx, Karl. *Grundrisse: Foundations of the Critique of Political Economy*. Translated by Martin Nicolaus. New York: Penguin, 1973.

Marx, Karl. "Theories of Surplus Value." In *The Collected Works of Karl Marx and Friedrich Engels*, 33:253–371. London: Lawrence and Wishart, 2010.

Mathews, Harry, and Georges Perec. "Roussel and Venice: Outline of a Melancholic Geography." In *Immeasurable Distances: The Collected Essays*, by Harry Mathews, 83–107. Venice, CA: Lapis Press, 1991.

Mattelart, Armand. "For a Class Analysis of Communication." In *Communication and Class Struggle*, vol. 1, *Capitalism, Imperialism*, edited by Armand Mattelart and Seth Siegelaub, 23–70. New York: International General, 1979.

Mattelart, Armand. "Modern Communication Technologies and New Facets of Cultural Imperialism." *Instant Research on Peace and Violence* 3, no. 1 (1973): 9–26.

Mattelart, Armand. *Networking the World, 1794–2000*. Translated by Liz Carey-Libbrecht and James A. Cohen. Minneapolis: University of Minnesota Press, 2000.

Mattelart, Armand. *Transnationals and the Third World*. South Hadley, MA: Bergin & Garvey, 1983.

McAfee, Andrew, and Erik Brynjolfsson. *Machine, Platform, Crowd: Harnessing Our Digital Future*. New York: W.W. Norton, 2017.

McLuhan, Marshall. "Interview with Gerald Emanuel Stearn." In *McLuhan, Hot and Cool*, edited by Gerald Emanuel Stearn, 266–302. New York: Dial Press, 1967.

McLuhan, Marshall. Introduction to *Explorations in Communication*, edited by Marshall McLuhan and Edmund Carpenter. Boston: Beacon Press, 1960.

Merton, Thomas. *Cables to the Ace: Familiar Liturgies of Misunderstanding*. New York: New Directions, 1968.

Merton, Thomas. *The Hidden Ground of Love: Letters*. Edited by William H. Shannon. New York: Farrar, Straus, Giroux, 1985.

Merton, Thomas. *Turning toward the World: The Pivotal Years*. Edited by Victor A. Kramer. San Francisco: Harper San Francisco, 1998.

Merwin, W.S. *The Miner's Pale Children: Prose*. New York: Atheneum, 1969.

Metcalf, Paul. *Collected Works*. Vol. 1, *1956–1976*. Minneapolis: Coffee House Press, 1996.

Metcalf, Paul. *Where Do You Put The Horse?* New York: W.W. Norton, 1986.

Miller, Stephen Paul. *The Seventies Now: Culture as Surveillance*. Durham, NC: Duke University Press, 1999.

Mitcham, Carl, and Robert Mackey, eds. *Philosophy and Technology: Readings in the Philosophical Problems of Technology*. New York: Free Press, 1972.

Mohawk, John. "Our Strategy for Survival." In *Basic Call to Consciousness*, 119–125. Summertown, TN: Native Voices, 1978.

Mohawk, John. "Technology Is the Enemy." *Akwesasne Notes* 1979, 19–21.

Morrison, Toni. *Sula*. New York: Alfred A. Knopf, 1973.

Motte, Warren F. "Eradications." *Romance Notes* 29, no. 1 (1988): 29–37.

Mumford, Lewis. *Technics and Civilization*. Chicago: University of Chicago Press, 2010.

Naipaul, Shiva. *Journey to Nowhere: A New World Tragedy*. New York: Penguin, 1982.

Nancy, Jean-Luc. *After Fukushima: The Equivalence of Catastrophes.* Translated by Charlotte Mandell. New York: Fordham University Press, 2014.

Newton, Huey P. *Revolutionary Suicide.* New York: Penguin, 2009.

"Not Good Germans." *Fifth Estate,* April 22–28, 1971, 3.

Paasonen, Susana. "From Cybernation to Feminization: Firestone and Cyberfeminism." In *Further Adventures of the Dialectic of Sex: Critical Essays on Shulamith Firestone,* edited by Mandy Merck and Stella Sandford, 61–84. London: Palgrave MacMillan, 2010.

paperson, la. *A Third University Is Possible.* Minneapolis: University of Minnesota Press, 2017.

Patterson, Orlando. *The Sociology of Slavery: An Analysis of the Origins, Development, and Structure of Negro Slave Society in Jamaica.* Teaneck, NJ: Farleigh Dickinson University Press, 1967.

Pettman, Dominic. "The Species without Qualities: Critical Media Theory and the Posthumanities." *b2o: the online community of the* boundary 2 *editorial collective.* April 23, 2019. http://www.boundary2.org/2019/04/the-species-without-qualities-critical-media-theory-and-the-posthumanities/.

Pontalis, J.-B. *Love of Beginnings.* Translated by James Greene with Marie-Christine Réguis. London: Free Association Books, 1993.

Postone, Moishe. "Necessity, Labor, and Time: A Reinterpretation of the Marxian Critique of Capitalism." *Social Research* 45, no. 4 (1978): 739–788.

Povinelli, Elizabeth A. *Geontologies: A Requiem to Late Liberalism.* Durham, NC: Duke University Press, 2015.

Power, Nina. "Toward a Cybernetic Communism: The Technology of the Anti-Family." In *Further Adventures of the Dialectic of Sex: Critical Essays on Shulamith Firestone,* edited by Mandy Merck and Stella Sandford, 143–162. London: Palgrave MacMillan, 2010.

Proyect, Louis. "Robert Brenner, Vivek Chibber, and the 'Organization Question.'" *Louis Proyect: The Unrepentant Marxist,* June 25, 2018. https://louisproyect.org/2018/06/25/robert-brenner-vivek-chibber-and-the-organization-question/.

Purves, Alan C. "Life, Death, and the Humanities." *College English* 31, no. 6 (1970): 558–564.

Pynchon, Thomas. *Gravity's Rainbow.* New York: Penguin, 2006.

Pynchon, Thomas. "Is It O.K. to Be a Luddite?" *New York Times Book Review,* October 28, 1984, 40–41.

Rand, Ayn. "The Anti-Industrial Revolution." *Objectivist* 10 (1971): 1–7.

Revard, Carter. "How to Make a N.U.D.E. (New Utopian Dictionary of English)." *Annals of the New York Academy of Sciences* 211, no. 1 (1973): 91–98.

Revard, Carter. "On the Computability of Certain Monsters in Noah's Ark: Using Computers to Study *Webster's Seventh New Collegiate Dictionary* and *The New Merriam-Webster Pocket Dictionary.*" In *Proceedings of the 1968 ACM [Association for Computing Machinery] National Conference,* 807–813. New York: ACM, 1968.

Rid, Thomas. *Rise of the Machines: A Cybernetic History.* New York: W.W. Norton, 2017.

Rosen, Kenneth, ed. *Voices of the Rainbow: Contemporary Poetry by Native Americans.* New York: Viking Press, 1975.

Ross, Kristin. *May '68 and Its Afterlives*. Chicago: University of Chicago Press, 2002.

Roszak, Theodore. *The Making of a Counter Culture: Reflections on the Technocratic Society and Its Youthful Opposition*. Garden City, NY: Anchor Books, 1969.

Russ, Joanna. Books. *Fantasy & Science Fiction*, November 1971, 18–23.

Russ, Joanna. Books. *Fantasy & Science Fiction*, March 1975, 39–45.

Russ, Joanna. "SF and Technology as Mystification." *Science Fiction Studies* 5, no. 3 (1978): 250–260.

Russ, Joanna. "The Wearing Out of Genre Materials." *College English* 33, no. 1 (1971): 46–54.

Russ, Joanna. *We Who Are About To . . .* Middletown, CT: Wesleyan University Press, 2005.

Sadako, Kurihara. *When We Say "Hiroshima": Selected Poems*. Ann Arbor: University of Michigan Press, 1999.

Sakaki, Nanao. "Memorandum." *Bombay Gin* 1979, 116–117.

Sanchez, Ricardo. "calles y callejones & memories." *A: A Journal of Contemporary Literature* 4, no. 1 (1979): 11–13.

Sassower, Raphael. *Technoscientific Angst: Ethics and Responsibility*. Minneapolis: University of Minnesota Press, 1997.

Schiller, Herbert I. "Freedom from the 'Free Flow.'" *Journal of Communication* 24, no. 1 (1974): 110–117.

Scholes, Robert. *Fabulation and Metafiction*. Urbana: University of Illinois Press, 1979.

Schumacher, E.F. *Small Is Beautiful: Economics as If People Mattered*. New York: Harper & Row, 1973.

Seidensticker, William D. "Language as Communication: A Criticism." *Southwestern Journal of Philosophy* 2, no. 3 (1971): 31–39.

Sennett, Richard. *The Hidden Injuries of Class*. Cambridge: Cambridge University Press, 1972.

Sheldon, Rebekah. *The Child to Come: Life after the Human Catastrophe*. Minneapolis: University of Minnesota Press, 2016.

Shenker, Israel. "Language Forum Hears Protests." *New York Times*, December 28, 1968, 18.

Shotwell, Alexis. *Against Purity: Living Ethically in Compromised Times*. Minneapolis: University of Minnesota Press, 2016.

Silver, David. "Looking Backwards, Looking Forward: Cyberculture Studies 1990–2000." In *Web.Studies: Rewiring Media Studies for the Digital Age*, edited by David Gauntlett, 19–30. London: Arnold, 2000.

Snyder, Terri L. *The Power to Die: Slavery and Suicide in British North America*. Chicago: University of Chicago Press, 2015.

Spanos, William V. "Breaking the Circle: Hermeneutics as Dis-closure." *Boundary 2* 5, no. 2 (1977): 421–460.

Spillers, Hortense J. *Black, White, and in Color: Essays on American Literature and Culture*. Chicago: University of Chicago Press, 2003.

Stevenson, Adlai E. *The Papers of Adlai E. Stevenson*. Vol. 8. Edited by Walter Johnson. Boston: Little, Brown, 1979.

Szent-Györgyi, Albert. *What Next?!* New York: Philosophical Library, 1971.

Talk of the Town. *The New Yorker*, July 4, 1964.

Thompson, E.P. *The Making of the English Working Class*. New York: Vintage Books, 1963.

Tiffany, Daniel. *Toy Medium: Materialism and the Modern Lyric*. Berkeley: University of California Press, 2000.

Tirella, Joseph. *Tomorrow-Land: The 1964–65 World's Fair and the Transformation of America*. Guilford, CT: Lyons Press, 2014.

Trump, Donald J. With Steve Bannon and Steven Miller. "The Inaugural Address." Speech, Washington, DC, January 20, 2017. https://www.whitehouse.gov/briefings-statements/the-inaugural-address/.

Turner, Fred. *From Counterculture to Cyberculture: Stewart Brand, the Whole Earth Network, and the Rise of Digital Utopianism*. Chicago: University of Chicago Press, 2006.

Walcott, Derek. "The Muse of History." In *Is Massa Day Dead?*, edited by Orde Coombs, 1–27. New York: Doubleday Anchor, 1974.

Ward, Barbara. *Spaceship Earth*. New York: Columbia University Press, 1966.

Warren, James Perrin. "'Catching the Sign': Catalogue Rhetoric in 'The Sleepers.'" *Walt Whitman Quarterly Review* 5 (1987): 16–34.

Weinberg, Alvin M. "Can Technology Replace Social Engineering?" *Bulletin of the Atomic Scientists* 22 (1966): 4–8.

Wiener, Norbert. *Cybernetics, or Control and Communication in the Animal and the Machine*. Cambridge, MA: MIT Press, 1948.

Wiener, Norbert. "A Scientist Rebels." *The Atlantic*, January 1947, 46.

Wilks, Yorick A., Brian M. Slator, and Louise M. Guthrie. *Electric Words: Dictionaries, Computers, and Meanings*. Cambridge, MA: MIT Press, 1996.

Williams, Raymond. *The Country and the City*. New York: Oxford University Press, 1975.

Windham, Lane. *Knocking on Labor's Door: Union Organizing in the 1970s and the Roots of a New Economic Divide*. Chapel Hill: University of North Carolina Press, 2017.

Winner, Langdon. *Autonomous Technology: Technics-out-of-Control as a Theme in Political Thought*. Cambridge, MA: MIT Press, 1977.

Winner, Langdon. "Peter Pan in Cyberspace: *Wired* Magazine's Political Vision." *Educom Review*, May/June 1995, 18–20.

Winslow, Cal. "Overview: The Rebellion from Below, 1965–1981." In *Rebel Rank and File: Labor Militancy and Revolt from Below during the Long 1970s*, edited by Aaron Brenner, Robert Brenner, and Cal Winslow, 1–36. New York: Verso, 2010.

Wong, Nellie. "On the Crevices of Anger." *Conditions* 1978, 52–61.

Zerzan, John, and Paula Zerzan. "Who Killed Ned Ludd?" *Fifth Estate*, April 1976, 6–7, 15.

Ziarek, Krzysztof. "The Way of the World, or the Critique of Cosmo-Technics." *Parallax* 16, no. 4 (2010): 3–15.

Zielinski, Siegfried. . . . *After the Media*. Minneapolis: University of Minnesota Press, 2013.

Zielinski, Siegfried. *Deep Time of the Media: Toward an Archaeology of Hearing and Seeing by Technical Means*. Cambridge, MA: MIT Press, 2006.

Zwigenberg, Ran. "Never Again: Hiroshima, Auschwitz and the Politics of Commemoration." *Asia-Pacific Journal* 13, no. 3 (2015): 1–22.

PERMISSIONS

Permission to reprint has been granted by the following:

The title line from the essay "The Master's Tools Will Never Dismantle the Master's House" from *Sister Outsider: Essays and Speeches by Audre Lorde*, copyright 1984 by Audre Lorde, copyright renewed 2007 by Audre Lorde, as well as quotations from the review *"Sula / A Review"* from *Amazon Quarterly* 2. 3 (1974) by Audre Lorde appear with the permission of the Charlotte Sheedy Literary Agency. Any third-party use of this material, outside of this publication, is prohibited. Interested parties must apply directly to the Charlotte Sheedy Literary Agency for permission.

Quotations from "For Each of You" are from *From a Land Where Other People Live* by Audre Lorde, copyright 1973 by Audre Lorde, and appear with the permission of W. W. Norton & Company, Inc. Any third-party use of this material, outside of this publication, is prohibited. Interested parties must apply directly to W. W. Norton & Company, Inc., for permission.

Quotations from "Getting Across" are from *Winning the Dust Bowl* by Carter Revard, copyright 2001 by the Arizona Board of Regents, and appear with the permission of the University of Arizona Press. Any third-party use of this material, outside of this publication, is prohibited. Interested parties must apply directly to the University of Arizona Press for permission.

Quotations from *Robert A. Heinlein: America as Science Fiction* by H. Bruce Franklin, copyright 1980 by Oxford University Press, Inc., are used by the permission of Oxford Publishing Limited. Any third-party use of this material, outside of this publication, is prohibited. Interested parties must apply directly to Oxford Publishing Limited for permission.

Quotations from "Routine for a Stand-Up Comedian" are from *Late Poems 1968–1993: Attitudinizing Verse-Wise, while Fending for One's Selph, and in a Style Somewhat Artificially Colloquia* by Kenneth Burke, edited by Julie Whitaker and David Blakesley, copyright 2005 by University of South Carolina, and appear with the permission of the Kenneth Burke Literary Trust. Any third-party use of this material, outside of this publication, is prohibited. Interested parties must apply directly to the Kenneth Burke Literary Trust for permission.

INDEX